The Universe

www.pocketessentials.com

Throughout this work, I am assuming that most scientists most of the time are rigorous in attempting to produce clear knowledge based on sound mathematical principles, and I cannot be held legally or morally responsible for outcomes, conclusions or facts that are proven to be untrue, irrelevant or simply wrong.

The Universe

Explained, Condensed and Exploded

RICHARD OSBORNE

POCKET ESSENTIALS

This edition published in Great Britain in 2007 by Pocket Essentials,
PO Box 394, Harpenden, Herts, AL5 1XJ, UK

A CIP catalogue record for this book is available from the British Library.

ISBN 10: 1 904048 82 X
ISBN 13: 978 1 904048 82 4

2 4 6 8 10 9 7 5 3 1

Typeset by Avocet Typeset, Chilton, Aylesbury, Bucks
Printed and bound in Spain

For Harriet, Helen, the Havenhands
and my Chemistry A level teacher whose
name I've forgotten.

Contents

The Universe

Introduction
The Beginnings of Cosmology

Somewhere out past The Venus Love Bar there is a notice that reads 'Last fuel before the end of the Universe' and, on the back of it as you go past, it says 'Last fuel at the start of the Universe', and as both could be true, there we have the conundrum. How does one define a beginning and an end in something that could possibly be limitless, or could be expanding, or might well bend back on itself? The latest theory suggests that the Universe is still expanding, like a nice big balloon, but then it might contract again into a much smaller thing. (Although we're talking pretty big spaces here.) The trouble is it's not just the spaces that are a bugger. It turns out that time mightn't be quite what we thought (and you have to be quick to get that one). Whichever way you look at it, the great spaces and vast distances of our galaxy alone are enough to bend the mind, and our galaxy, it turns out, is just one of thousands, or millions. Trying to think about what we call the Universe means trying to think about everything that might have existed, and anything else that might also come into being and what might exist in the future. As Winnie the Pooh once pointed out, this does make your brain ache. Are there parking meters at the end of the Universe, and what time limit is there on the meter? This is the stuff that we all worry about and it reflects our basic human desire to know about the Universe, and also to try and grasp it on a human scale. 'There's nowt so queer as the Universe,' as some famous Northerner once said. Interestingly, the more we learn about the Universe, the more peculiar it seems to get. This doesn't stop people, or theoretical physicists anyway, from trying to develop a unified theory of everything. This is like trying to establish some basic principle, or set of rules, that will describe everything in the Universe forever. There are those who suggest that this might be a little bit over-ambitious, but we'll consider that question later (along with the question of black holes and the no-boundary proposal). Here we are just going to look at exactly why the Universe is such a problem, and why we worry about it (if we do).

This is another funny thing. Here we are wafting around in the middle of nowhere and we try to make the Universe fit our bug-eyed, small-brained view of everything. To put that scientifically, we might say that 'man is the measure of all things' and that our view of the earth, the planets and the stars has always been limited by our humanness. We have always been convinced that we are the centre of the Universe, the key factor in everything. To suggest otherwise has, throughout the ages, been seen as ridiculous, illogical and generally treasonable or heretical, or both. It's obvious, surely, that the moon, the sun and the stars circle the earth, which must therefore be at the centre of the Universe and everything goes around it in nice circles. God ordained all this and it all works perfectly, so there is no sensible reason to doubt what we see with our own eyes. That, at least, was the general line of argument for a couple of thousand years. This idea has, of course, changed in the last few centuries as technology has allowed us to see, and hear and record, many more things than we ever dreamed of. Our little human view of the Universe is being blasted apart all over the place. Without a doubt the telescope is the single most revolutionary bit of technology we have ever dreamed up and, with it, our views of the Universe began to change dramatically. That was where Copernicus and Galileo came in, pointing out to everyone that the earth went around the sun and not the other way around. There was actually hardly any reaction when Copernicus first said it, and almost no widespread reaction for 50 to 100 years. This was a very quiet revolution indeed. Now we look at the stars and we think, 'The light from that place definitely took 500 years to get here,' because we all now know that galaxies go on forever and we last for little specks of time. Some people's reaction to the real difficulties involved in thinking about all of this stuff is to drink beer and watch football, or to take up knitting, and try not to worry about it. This is a very human reaction but everyone, at some point, has to have a quiet worry about it all and that's what we're doing here. We're taking a look at the Universe for Simpsonites (and if the Universe could be as funny as *The Simpsons*, that would probably be a good thing). Indeed, the episode where Homer Simpson gets to grips with the various natures of different realities is the best bit of scientific popularising in existence. The questions Homer may well have put are, 'Where are all those aliens?', 'Are they boring?' and 'Do they drink beer?' These are very important questions as it's really a way of saying, 'Are aliens like us?' If they're not, it could be difficult to get on with them, since we're not very good at getting on even with our own types. Conceptually speaking, it's all about thinking outside the box, thinking in a way that is critical rather than commonsensical. Imagining the Universe and what is in it is the really hard bit for all of us. It's yoga for the brain. As Einstein once said, 'I'm going sailing.'

Anyway, first of all, what do we mean by the Universe? To quote the *Oxford English Dictionary*, the Universe is, 'The whole of existing or created things regarded collectively, all things, including the earth, the heavens, and all that is in them, considered as constituting a systematic whole'. When astronomers talk about the Universe, they mean everything that is accessible to our observations, but that keeps expanding, as does the way that all of these things interact. The Universe includes all that we can survey or experiment on, from the moon that orbits our own planet out to the most distant islands of stars in the vastness of space. We also assume that the Universe is all joined up, and made of the same sort of things. Since we cannot visit most of the Universe, we rely on the information it can send to us. Fortunately, we receive an enormous amount of cosmic information all the time, coded into the waves of light and other forms of energy that come to us from objects, stars and galaxies at all distances. Now that we have learnt, or theoretical physicists have, to decipher all this information, we can seriously analyse the Universe. So the main task of astronomy is to decode all that information and assemble a coherent picture of the cosmos. We could say that at the end of the day it all comes down to how we observe the Universe and what thinking about those observations does for our ideas of science, society and self. The Universe ought to inspire a deep sense of awe in everyone, but instead we block it out with hideous orange street-lamps so that we can't see the stars.

As the evidence accumulates about the nature of the Big Bang, you would think that we would more and more adopt scientific attitudes to the world but, in fact, religion seems to be on the march everywhere. That we know what makes the Universe tick would, you would think, help us to understand everything else better but, in a strange way, it seems to scare people. It's almost as though we want to stick with the idea that the world is flat and supported on the backs of elephants or tortoises, although that's a bit like believing that David Beckham is as important as Stephen Hawking. We call this 'ideology versus science' and, unfortunately, the myth-makers, who are often the media, frequently have the upper hand and promote ideology. Take, for example, the environmental debate about global warming. Almost all scientists believe that the evidence is overwhelming yet the media constantly portray the issue as speculative. Politicians encourage this, as they don't want to do anything about it. So many people – politicians, religious bigots, big companies and tobacco companies – have an ongoing interest in de-bunking science so that superstition rules the airwaves. (Remember the so-called debate about the link between smoking and ill-health?) But at the same time, we know more and more scientifically and seem unable to do anything with it, except make weird science fiction movies. This poses what we can call the Universe gap, between what we know and how we act.

Perhaps we are alone in the Universe, rattling about in gigantic spaces with just dust and rubbish for companions, or there may be lots of other civilisations out there just waiting for us to get in touch. Apparently, loads of Americans have been abducted by aliens but they were warned to keep quiet, or at least only to go on a chat show and talk about it. How we get in touch with all of these other civilisations, or even vaguely establish whether they exist, is one of the trickier questions that we face. Given that we now know that there are thousands of planets out there in various galaxies that could support life, it would seem credible that other life forms may have evolved. Or to put that another way, doesn't it seem a bit improbable that we lousy humans are the only intelligent life form in all of these endless voids and galaxies? It may seem improbable, but all the information we have at the moment (and by that we mean hard, scientific evidence) suggests that this may be the case. We now listen out all the time for messages from space and we are able to monitor vast tracts of the Universe, but we haven't yet had a single hello from anybody, or at least verifiable hellos that don't involve Elvis Presley or the Scientologists or whoever. This is rather odd in one sense. Logically, one might surmise that there must be other life forms out there but perhaps they are just too far away, or of such a different form that communication is, for the moment, impossible. However, given that the Universe is much the same from one end to the other, in terms of what it is made of and the way it works in terms of gases, radiation, light and movement, you would think that scientific development in other civilisations would have to be pretty similar to ours, and so they should come up with similar modes of communication. Thus, logically, if they are out there looking at the stars like us, but from another planet, they ought to be able to notice the peculiar things that go on, on earth. Indeed, perhaps that is why they haven't been in touch; they don't want to get involved with such a destructive bunch of lunatics. However, it is just as possible that we are the only life form in the entire Universe, brought into being by a series of accidents, fusions, natural selection and sheer improbability, and the way that many humans behave suggests that this could be the case. It is extraordinary, however, how we mere humans have so relentlessly over the centuries tried to work out what makes the Universe tick and have been able to discover so much in scientific terms about our planet, our Universe, and our physical being. It is just as extraordinary that we seem to have discovered so little about our social being, and to be able (simultaneously) to explore the stars and to fight like dogs in the gutter. It seems as though every time someone makes a scientific discovery, someone else dreams up a particularly appalling use for it, like atomic bombs or car technology that is destroying the earth's atmosphere. One recent argument for space travel is that we are messing up the planet so quickly that we will need to get off en masse within the next couple of hundred years. This is what is known as curious logic.

Before we get to the end of the planet, it is worth going back to one of the original philosophical questions, 'Where did everything come from?' This can properly be described as the grand-daddy (or grand-mother) of all questions and it can rightly be seen as what distinguishes us from other, non-rational beings. Rather than just reacting to our environment, as many animal species do, often in intelligent ways, we as a species have constantly tried to work out where our planet came from, why things happen as they do and what causes such events. The earliest explanations tended to be mythical, or religious, but gradually, particularly through observation, observers started to pose questions that could really be described as philosophical or scientific. As far as we can tell, the early Greeks brought these things to fruition and produced all sorts of fascinating ideas that got Western science and civilisation off to a head-start. Greek philosophers more or less developed the scientific method of thinking 'What if...?' or of just thinking about the Universe, both in abstract terms and in the sense of thinking about measuring it, which is effectively where it all begins. Their idea of 'natural philosophy' was to think about everything in terms of its inter-connectedness, or of systems, and to speculate on how one bit affected another. It is still quite staggering to see what they came up with, based on a few measurements, some mathematics and a great deal of pre-computer intelligence. From the invention of writing, about 5,000 years ago, to the Greeks speculating about the atomic structure of the Universe, was a period so short that it was like a fruit fly developing language and inventing the computer before it died in two weeks (or however long they live). There are other very interesting questions about why Greek civilisation suddenly blossomed in the way that it did, and also then declined, but that's a whole other debate.

As I've said, we are confronted with the slightly difficult questions of 'Where did the Universe come from?' and 'What is it made of?' Perhaps God made the Universe, or a group of gods who have since fallen out, which would explain all the war and pestilence. Or perhaps the Universe just happened, by mistake, and we were part of that mistake, and later we'll find out why. Basically, for all of us who are not theoretical physicists, or God, we have a problem in understanding what all of this stuff is about and that is why we need to think about it in non-mathematical ways. Already you are asking what is this talk about the Universe and understanding it in ways that are non-mathematical, and that is a good question. As we philosophers like to say, a good question is where it all starts from; where it goes from there is anybody's guess. The point is that most of the development of science and astronomy has been bound up with the development of mathematics (or numbers as we non-professionals like to say). The question of where numbers come from or who invented them is another one of those rather tricky questions that they don't teach you about in school. Pythagoras and his Greek mates were all convinced that numbers were mystical and spiritual, sort of alive, and made up the way the Universe functioned. This is a very long way from the origin of mathematics in counting things, like sheep, cows and eggs; but without mathematics, none of the scientific developments would really have been possible. The development from numbering things to being able to devise, and solve, mathematical problems is another one of those great mysteries that almost defies explanation but, without it, our ability to measure and analyse our observations of the Universe would be non-existent. Going with numbers meant that we could measure things by inference rather than flying to the moon and this opened up thought about the Universe in the most exciting way. It is a little-known fact, for example, that a Greek mathematician, one Eratosthenes, worked out the circumference of the earth almost entirely accurately in the third century BC. This was done by mathematical calculation and a bit of genius, but it showed two things: that it was an accepted idea that the world was round, and that astrology was seriously getting to grips with ideas of observation and development. He worked at the famous library in Alexandria, which was a world centre of learning at the time, but which naturally got burnt down with all the books (rolls actually) in it. We Europeans then invented the Dark Ages in which morbid Christianity replaced thought, but the Arabs kept the Greek knowledge alive, thank goodness.

One Aristarchus of Samos had actually said that the world was not the centre of the Universe back in the third century BC and Eratosthenes probably agreed with him, but they had no hard evidence to back it up. That came a couple of thousand years later. Aristarchus also tried to measure the distances to the sun and the moon, which was pretty ambitious. He used various mathematical formulae he had devised which weren't that accurate, but he gave it a good try. Aristarchus put out a snappy little book called *On the Dimensions and Distances of the Sun and Moon*, which was not, as far as we know, a best-seller, but it was his only work. He also estimated that the moon was a quarter of the size of the earth; not very good, but based on calculation, which was the important bit. To blithely move the centre of the Universe from the earth to the sun was so loony that absolutely no one paid any attention to him at the time. Samos was, in fact, an island seemingly full of crazy Greeks, including the grand old man of the hypotenuse, Pythagoras. For one small island to have so much influence on the development of science, cosmology and mathematics is almost as weird as the idea that we are the most intelligent life form in the Universe. Mathematics, as applied numbers and theorems, really is the most amazing tool in the scientist's kitbag and it all starts with Mr P, who also insisted that members of his cult did not eat beans. Pythagoras was both scientific and religious, but it is with him that mathematics really starts as the process of deductive reasoning, or saying that if a then b then possibly c. This practice of reasoning, mixed with speculation, is what led the Greeks to so many of their extraordinary ideas and explanations. Pythagoras believed that certain theorems were sacred and that 'all things are numbers', ideas that led him to form a sort of cult, but without this magic of numbers we might still be riding round on donkeys and looking for the edge of the flat world. That we found ways of saying, 'If I observe this star, then I can deduce the movement of the earth, because of the one's movement in relation to the other,' is what has allowed the entire development of astronomy and theoretical physics. Or to put that another way, scientific knowledge is based on observation, inference, deduction and speculation. Just how we developed, from a few Greek speculations to understanding the nature of matter, time and space, is a very complicated story, and a lot of it has happened in the last hundred years.

Our known Universe, as we like to call it, used to consist of the earth and the bits we could see in the sky, like stars, the moon, the sun and a few other things that flew about or disappeared, like clouds, meteorites, falling stars or eclipses. Thus, the Universe simply meant our little planet and the things that revolve around it. This was the original view of the Universe that the Sumerians, Babylonians, Indians, Chinese, Greeks and others had, mainly based on what they saw when they got up in the morning and the things they noticed before they went to bed very late at night. This is what I meant earlier on by saying that our view of the Universe was based on our humanness; it was what we could see and understand. So you have to think of some ancient Sumerian/Babylonian waking up in the early morning and watching the stars disappear over the horizon at one side and the sun coming up at the other. As a farmer who probably didn't read or write and who wasn't a member of Mensa, you'd pretty much assume that the stars went to bed and the sun came round on a daily basis, just in time, in fact, for the day's agricultural work. Most of what we call culture and science developed out of attempts to understand and predict what sorts of things were going to happen next, and how things like floods, storms and the weather generally affected everyday life. Thus, gods and nature got mixed up and religions were cobbled together to explain all the nasty things that might happen. The Maya civilisation was particularly good on sun gods and sacrificing people, a ritual that might make a comeback. We can roughly say that religions were used as a kind of mythological means to try to understand the innate nastiness of the world and to make sense of nature, thus providing a bit of reassurance in a frightening and unpredictable place. At the same time, astronomy tended to get tangled up with religion as well but, from the beginning, astronomy was based on actually watching the stars and recording what they did.

The Babylonians, who lived in what we now idly call Iraq (it may be renamed North Texas), inhabited what was one of the places where world civilisation got started, somewhere around 7,000 years ago. They inherited some ideas from the Sumerians but gave us legal codes, mathematics and astronomy, which have proved to be of supreme importance in the development of human knowledge. The Sumerians basically invented the writing and alphabet stuff along with a little astronomy, and the Babylonians added abstract thought and applied mathematics, as well as art, society and culture. Mind you, who started what, and when, can get to be a very complicated argument, but we can safely say that Mesopotamia contributed a lot to early human civilisation. In particular, the Babylonians divided the day up into 24 hours and the circle into 360 degrees. They also worked out a cycle of eclipses, allowing lunar eclipses to be predicted. We Westerners like to go on about the Greeks, but there was an awful lot that went on before Homer and his merry band started thinking. For example, the wonderful and lengthy poem *The Epic of Gilgamesh* originates with the Sumerians and is one of the oldest works of literature in the world, demonstrating that they had a sense of the human's flea-like place in the Universe. Like many early religions as well, they worshipped many gods and did so to try and appease the powerful forces of Nature and the Universe that affected them. The mysteries of the sun, the moon and the stars tied in with questions of life, death and the weather, which determined the course of everyday life. Interestingly as well, the Sumerians came up with the idea of the calendar which, when you think about it, is pretty significant in a farming society. Based on observation of the sun, moon and stars, it allowed people to predict what was going to happen next – always a useful bit of knowledge. The calendar is more or less the start of applied astronomy, of science and of man trying to understand and control the Universe. That sounds like rather a large claim but it is the case that it was through observation, and recording, that our understanding of the Universe began, very slowly but definitely, to develop. From what we know, the Chinese, the Babylonians, the Mayans and the Egyptians all did some serious looking at the sky and developed different methods of recording the information they obtained. The irony is that by simply looking, you got rather a strange picture of the Universe, one that was basically completely wrong. Hence, the expression 'to see it with your own eyes' is about as true as 'the stars only come out at night'. That could be rephrased as the stars come out at any time over the life of the Universe, expand and grow and then, if they feel like it, explode and become a supernova and then a black dwarf star that sucks in gravity. You can see why people liked to stick to the idea that the sun comes up everyday in the east and goes to bed in the west. It's a lot more reassuring.

Don't forget a lot of people thought for a long time that the earth was flat and held up by tortoises, whereas now we know that they were terrapins. Actually though, not that many people thought the earth was flat and from very early times it was realised that it was probably curved, from the simple observation that you saw the tops of ships first and then the whole thing. The sun and the moon seemed to go in circles as well and so the assumption of spheres got built into astronomical thinking from the very beginning.

What the sky and the Universe were, however, puzzled everyone, and many originally thought that the sky was the limit and, as a fixed entity, was probably solid, or at least an impenetrable barrier. The Egyptians, for example, thought that the sky was the body of the goddess Nut, and that the earth was the body of the god Qeb. Along with others, like the Polynesians and the Mexicans, they believed that celestial bodies were gods, ruling over us. The sun was obviously the source of all power and therefore thought of as an important god, particularly by the Incas. Independently, however, the Babylonians and the Chinese began just to observe the stars and to plot their movements and, like the Egyptians, to formulate basic calendars that reflected movements of the heavens and the seasons. What began to be realised from early on was that there was regularity in the Universe and thus the idea of a systematic whole, of a coherent Universe, began to emerge. It is important to stress, however, that our knowledge of what was generally known in these early periods is very sketchy and often reliant on later, and not terribly trustworthy, sources. Given that, for example, a vast number of the world's books were burnt in the great fire at Alexandria library, what remains could be a completely partial view. All cultures seem to develop some form of thinking about the nature of the Universe and, whether mystical or mechanical, civilisations seem to require some explanation of man's role. Religion and government tended to be mixed up and myths were part of the ruling civil forms, so historically this tended to limit scientific views.

However, from the evidence we do have, it seems that the quantum leap forward was down to the Greeks, who appear to have re-thought the entire Universe, dreamt up maths and philosophy and even had ideas about atomic structure. Sometime around the seventh century BC, the Greeks, who actually lived all over the Mediterranean, appear to have developed a view that the Universe was basically a rational place that followed natural and universal laws. This idea, that everything is inter-connected and rational, is essentially scientific, although the Greeks called it natural philosophy. One person, Thales of Miletus, is credited with pointing out that an eclipse of the sun was not a mystical thing but probably due to the movement of the planets – and bingo!, you have the beginnings of a scientific theory of the Universe. This was in the year 585 BC, which can be verified by astronomers and thereby provides the first clue of cosmological science.

At this time (and today) about 2,000 stars were visible to the naked eye and Thales was, apparently, a keen observer. A bit later, Aristotle summed it all up in his book *On the Heavens* (340 BC), another one of those snappy titles destined for the best-seller list. Aristotle was a great observer of things, an early scientist and a great system builder in philosophical terms. He embodied the Greek approach of critical inquiry and open speculation, as well as summarising the state of cosmology at the time. He argued that the earth was round, based on looking at the shadow of the earth on the moon at eclipses, which was also round. He also mentioned that the North Star appeared in a different position when viewed in the south, which would happen if the earth were spherical. He added the well-known fact of ships appearing slowly over the horizon to clinch things and then spoiled it by saying that the earth was stationary and everything else went round it.

However, as I have said before, Aristarchus of Samos

had also ventured the idea that the earth went round the sun, albeit in the same neat circles that Aristotle believed in. Both of them, however, believed in that speculative mode of thought that went, 'What if the earth turned around every day as well as going around the sun?'

The next theorist of the Universe, and the most important for a very long time, was Ptolemy, whose proper name was Claudius Ptolemaeus and who worked in Alexandria and died in AD 180. All he did was to draw together all of the existing knowledge of the Universe and to outline a complete system, what we call the Ptolemaic system, which dominated cosmological thinking for the next 1,400 years. In the currency of the day, Ptolemy was concerned with 'saving the appearances', which meant to make the description of the Universe fit what was visually observed. Not everyone was concerned with doing this, of course. The religious lot thought that spiritual essences were more important than empirical realities. In fact, this 'saving the appearances' made life difficult because there were some rather complicated things that could be seen, like eclipses, and the fact that some planets sometimes seemed to move backwards. In trying to construct a theory of the Universe in which all the stars and planets moved in neat spheres, Ptolemy had to produce some pretty tricky maths, which he did and which seemingly did explain everything. This was, and is, very impressive, particularly given how wrong it actually was. Basically, what Ptolemy said was that the earth was the centre of the Universe and that there were eight spheres that moved the moon, the sun, the stars and the five planets around. His system was pretty accurate for predicting what could be seen with the naked eye but it led to strange theories, like the fact that the moon supposedly sometimes came twice as close to earth as it did at other times. You might think that, if that were to happen, the moon would look bigger (or twice as big) but clearly it didn't, apart from when it was half-moon, quarter-moon and stuff like that. Also, there was the notion of the fixed outer sphere which somehow moved the stars around, kept them all in their place and formed the limit of the Universe, beyond which there was loads of empty space – and heaven and hell! This idea of sort of crystalline spheres that moved mechanically around was very attractive and, with Ptolemy's fancy mathematics, it did seem as though a complete explanation of the Universe was there. This was something of which feudal lords and Popes very much approved because it suggested a fixed, unchanging world in which their rule was never challenged. Ptolemy also estimated the distance to the edge of the Universe, which was the stars, and he came up with about 75 million miles, which is about 1,000 per cent wrong, but was quite bold and original.

Ptolemaic astronomy survives because his work was translated into Arabic, and was kept alive and transmitted to the West by Islamic scholars, who merged it with Aristotle's cosmology to give the view that accorded with the Christian outlook. Dante's great work *The Divine Comedy* reflected this fixed view of the Universe, in which the different spheres of heaven and hell descended to the centre of the earth in the four-teenth century, when order seemed paramount. Just over the horizon, however, was the rising star of Copernicus, who would remind the world of Aristarchus of Samos's wild claim that the earth went round the sun, and that it whizzed around on its axis every day. It was because Copernicus went on to prove that this was true, that it 'saved the appearances' and fitted the facts, that his work was to create a revolution in our ideas of the Universe.

There could not have been a more reluctant revolu-tionary than Nicolaus Copernicus, who merely set out to iron out the faults in the Ptolemaic system and to account for the oddities of the mathematics that Ptolemy used. As early as 1507, Copernicus wrote a short, hand-written book in which he put forward the idea that the earth-centred Universe was fiction, and that putting the sun in the centre would be more logi-cal, but he wasn't certain and only a few people read his sketch called *The Commentary*. Copernicus spent many further years working on the mathematics, and also developing the idea that most of the strange things that could be observed in the Universe could be explained by the earth's motion. Using more of his own observations, Copernicus developed his arguments over the next 30 years and put his final position in a book which was to revolutionise our understanding, which he called *De Revolutionibus Orbium Coelestium (Concerning the Revolutions of the Heavenly Orbs)*. He was reluctant to publish because he didn't want people thinking he was a nutter, and somebody did later write a play taking the mickey out of the loony who suggested things didn't go around the earth. Everyone urged him to publish and he kept saying it wasn't ready and prevaricating so, of course, by the time the book was actually printed, he had died, never having seen the finished thing. Thus, the book's publication and his death in 1543 has to be seen as the most important date in the development of our understanding of the Universe, but it was such a quiet revolution that it was practically another 100 years before even most scientists totally accepted it.

The fact that other astronomers, and particularly Galileo, set out to prove all of the things that Copernicus claimed helped a lot, but it was still an uphill struggle, with perhaps only a couple of dozen people agreeing with Copernicus even 100 years after his death. Galileo, a whiz at astronomy and self-promotion, used the newly invented telescope to show that the Universe was quite different to the old fashioned fixed notion and that there were lots of things moving about out there that could only be explained in terms of Copernicus's newfangled ideas. Not unnaturally, the Pope wasn't having any of this radical nonsense and had Galileo put under house arrest for the rest of his life. This proved two things: one, that new technology was the way to develop science, and two, that there would always be resistance to ideas that suggested that the Universe was more complicated than we previously thought. Over the next few hundred years, this would happen again and again. Unfortunately for the general reader, the more we learn about the Universe, the more complicated it seems to get, but on another level, the more entertaining.

From Stars in their Eyes to
Telescopes and Beyond

Our modern ideas of the Universe, by which we mean basically scientific ideas that depend on observation, explanation and evidence, can be traced back to Nicolaus Copernicus in the fifteenth century, but they needed an awful lot of development before we were up to speed. Working out which way the planets go around isn't quite the same as working out when the Universe started, or how, or of understanding the forces at work in the Universe. The reason that Copernicus was so important, however, was that his model of the Universe was the first that was entirely based on a rational examination of all the known facts. In other words, he didn't let belief or general opinion sway him in any way. In one very interesting sense, his model was based on pure theoretical speculation rather than common-sense views: it got to the heart of the fact that the Universe wasn't like a bigger version of human society but had its own strange laws. The older Ptolemaic (earth-centred) system had used fancy maths to bend our view of the Universe into something that just about worked, as long as you ignored the bits that didn't add up, like the fact that some planets appeared to go backwards and have funny orbits. The continued belief in the earth-centred theory was undoubtedly because everybody wanted there to be a small, regular Universe, in which everything moved in neat spheres with the earth in the middle. Copernicus's model went against this common sense, which is why he didn't want to publish it, and, because of that counter-intuitive approach, science has never looked back. Copernicus didn't prove that the Universe was a lot more complicated than everyone thought, but he pushed the door open so wide it couldn't be shut again.

Once again, it would be actual observations that would take the arguments further, in the persons of Tycho Brahe and Johannes Kepler, who teamed up in Prague in 1601 to further their mutual interest in astronomy. However, it was always going to be observations teamed with mathematical theory that would expand true knowledge of the Universe. Brahe actually believed in the earth-centred stuff but couldn't help using his eyes. Kepler was, by all accounts, a lousy teacher, never in good health and couldn't see very well, which is why it was such a good thing that Tycho Brahe was such an unbelievably good observer of the heavens. Tycho's incredibly accurate observations, particularly of Mars, even without a telescope, threw up lots of problems with the basic Ptolemaic theory, in that his new observations didn't fit properly with the theory. Statistically speaking, Brahe's observations were estimated to be ten times more accurate than anybody else's at the time. He also discovered a supernova in a constellation of Cassiopeia, which is pretty good going for the time. (The instruments can still be seen in the Prague Museum of Technology.) Brahe himself believed in the Ptolemaic view of the Universe, which was ironic since his work effectively unpicked it. Apparently, he thought that God must have created a perfect Universe, in which everything would be regularly and neatly ordered and that, being a very tidy god, he would not have put stars randomly all over the place in huge empty spaces. (Because God only does rational, organised things by definition.) It is interesting just how often early theorists of the Universe start from a philosophical position and impose that belief on their observations. Brahe was a bit of a Neoplatonist (believed in essences and perfect structures), and so would then attempt to make the facts fit the theory, rather than consider them with a completely open mind.

Fortunately, Kepler wasn't like that and kept worrying about the problems that Tycho's observations had thrown up. He was also the maths whiz kid and, after Brahe died, he inherited his position of Imperial Mathematician (which sounds like something out of Harry Potter) but he couldn't manage to get paid – so not much has changed there for scientists and researchers. Kepler used Brahe's observations to show that the path of the planets was not circular, but actually elliptical and then he apparently cracked the only joke of his entire life when he said, 'I've laid an enormous egg' (you have to think about it). Kepler developed what he called laws of planetary motion, which again were quite extraordinary in terms of thinking about the relationship of the planets to the sun, and he alluded to some strange force he speculated about. He called it a 'whirling force' and it clearly prefigures what we later called gravity. His first two laws were published in 1609 and they were called the New Astronomy, for obvious reasons. One of Kepler's final three laws predicted that the way a planet behaved was in relation to its distance from the sun, which, being the largest planet, clearly affected the others, which also explained the elliptical orbits. He also calculated the most exact astronomical tables so far known, whose accuracy turned out to be right out there with the later observations made with telescopes.

Kepler's laws, which we shan't go into in any detail, dealt with the regularities of the movements of the planets and his whole outlook was summarised in the title of his great work, *The Harmony of the World* (1619). It would be fair to say that Kepler's use of mathematics to think about the motions of the planets laid the entire framework for conceptualising how the mysterious forces of the Universe functioned. Without his planetary laws,

Newton's theory of gravity would not have been possible nor any of the new cosmology of the eighteenth and nineteenth centuries. Kepler unfortunately died when travelling across Europe to try and get paid for some of his earlier work. In keeping with the earlier tradition of Copernicus, Kepler carefully predicted that Mercury would cross the sun on 7 November 1631 and then died just before it happened. The former demonstrated the regularity in the behaviour of the Universe that Kepler so admired. Another ironic regularity was that Kepler started out in life a pauper, worked incredibly hard and then died a pauper chasing up his meagre pay.

Kepler established that there was a connection between the time a planet took to go around (in its egg-shaped orbit) and its distance from the sun. This was important stuff, as was the utterly bizarre suggestion by the 'mathematician-comic' that it might be possible, working backwards, to work out the moment of the creation of the Universe. Now that was like laying a dinosaur egg and telling the world's best joke all at once. Admittedly, Kepler came to this idea because he thought the Universe had all of these strange and beautiful harmonies, sort of musical harmonies, but however you get to it, it is a profound and wonderful idea. Like Pythagoras before him, Kepler had believed that there was a mathematical and musical harmony at the heart of the Universe and that we might be able to unlock it, to find the hand of God. It is a mystical idea but the attraction of perfection is clearly a powerful one, albeit not up to the scale of gravity, which holds the Universe together in a more mundane but efficient way.

In 1600, when Kepler and Brahe first met, most people still believed that the earth was the centre of the Universe and that things moved about in perfect spheres. By 1700, hardly anybody believed this and scientific explanations were all the rage. The most curious thing about these developments in knowledge is not only their inter-connection, but also the strange and unlikely way in which advances in human knowledge get made; sometimes it is a miracle that any get made at all. For every astronomer in the sixteenth century, there were at least half a million others dedicated to superstition, war, alchemy and conquering other countries with extreme prejudice. Kepler really made the break between mysticism and astronomy, and made scientific explanations of planetary forces inevitable. One of the intriguing questions Kepler's endless puzzling away in a darkened room produces, apart from the question of why someone would spend ten years looking at the same observations, is just why is mathematics such an accurate and powerful means of speculating about the Universe? Or, in other words, why does mathematics work?

As though just to disprove the theory that astronomers were all boring, along came Galileo, cosmology's answer to Frank Sinatra – sort of Italian and sort of smoochy and a lot of, 'I did it my way'. Actually, he was completely Italian and generally a very good scientist. Born in Pisa in 1564, he is forever associated with the telescope, the Inquisition and the leaning tower from which he supposedly dropped things to prove that lighter and heavier things fall together, which is true. The bit about the tower may well have been a legend but he did experiment a lot on all sorts of things. He was also rude about lots of people, including the highly venerated philosopher, Aristotle, and charmed a lot of other people. Galileo got his hands on the newly invented telescope, improved it a bit and rushed into print, telling all astronomers to get one of these fancy new devices and look at Jupiter. He didn't actually claim to have invented the telescope but he definitely implied that his version of it was the real thing, which got him a well paid job via the Senate of Venice and thereby allowed him to do some very important stargazing. If he didn't write the song, he certainly exemplified the notion of 'I did it my way', because he pursued with great vigour a path that enabled him to make great discoveries, get rich and still be rude to people who were rather powerful. Rather more importantly, he also discovered Jupiter's satellites which, to coin a phrase, put the flying fox amongst the heavenly chickens, and he saw that there were far more stars than you could poke a celestial stick at. He also noticed that

Venus, like the moon, had a range of sizes and changes that would be impossible according to the Ptolemaic version of the Universe (because of Venus's supposed place between the earth and the sun). Jupiter's moons, which clearly orbited around said planet, proved that there was not one fixed centre to the Universe, i.e. Rome, and that therefore, as I said before, 'there's nowt so queer as the Universe'.

Strangely, at this very moment, a spaceship called Galileo, which has been orbiting Jupiter for the last eight years, providing untold information about Jupiter's moons, has just burnt up into the atmosphere after sending back evidence that there is water on one of the moons. Thus, 2003 may be the year when Galileo's observations pinpoint where there may indeed be life in the Universe. Back in 1610, Galileo made some speculations about these moons that he spied through his telescope, which sounded like a theory of gravity, but, importantly, it was the actual observations that changed things. Like, for example, his observation that our moon had craters and mountains, which meant that there had been change and development there as well as on earth, and thus once again contradicted the traditional view of a Universe fixed in aspic. The scientific implications of this were enormous, as were the political and religious effects. So, like a good diplomat, Galileo went around saying in print that the Church et al. were all backward and not up to speed, which led, as surely as the phases of the moon, to his getting sorted by the Inquisition.

Actually, the Papacy didn't really want to be terribly nasty to Galileo and Pope Urban VIII liked to have chats with him about the Universe and philosophy, but not in front of the children. However, when Galileo went for the big one and published a book in 1632, known as the *Dialogo*, which convincingly and completely set out the new Copernican system of the Universe, Pope Urban VIII lost it completely and condemned and prohibited all of the Copernican ideas. This sudden change of mind was very odd, as the Church almost seemed to be accepting the new view of the Universe, and caught everyone by surprise. Poor old Galileo had to write a retraction and sign it, and recant in public all his trendy ideas about the Universe, and then go off and hide in his villa. It's not known whether he blamed the newfangled telescope or if it was Venus that led him astray, but we know that his contribution blatantly outlasted that of the urbane Pope Urban (who sounds like a mass transit system).

What did this imbroglio prove? Mostly, that we all liked our fixed notions of the Universe, and that the Church very much liked the idea of its right to say what was right and what was wrong. But it also showed that science and religion are pretty much completely opposed and that this breach can be seen as the beginning of the end for the Holy Roman Empire. Perhaps that's what Pope Urban, belatedly, recognised. If the Church taught that the Universe was simple and perfect, and as Aristotle had argued, complete and full, then these newfangled ideas clearly implied that the Church was wrong. Given that the Pope was supposed to be infallible, this could be a serious problem. The idea that there were planets out there with moons flying about them that no one had ever known about, and that the earth whizzed about turning on its axis, was enough to turn a saint into a swearing sinner. Galileo had apparently called people who still believed in the old sun went round the earth stuff 'dumb idiots' and he implied the Pope was a simpleton in his *Dialogo,* so his retraction, as he well knew, looked pretty feeble. Galileo had seen through his telescope that the Milky Way was composed of a 'myriad of stars' and that nothing in the Universe was as simple as the earth-centred model. He also recognised political power, however, and bent his knee accordingly, with mockery and disdain.

Galileo also understood what he had seen through his telescope, however, and he knew that it was real and that nothing could put the Venus back into the Milo, or the Pope back into infallibility. He died in 1642 at the age of 78, quite sure that science would come out on top. Someone else once said that Galileo was the best scientist of the twentieth century, and the first one to use his eyes, or at least his telescope. Sometime earlier during Galileo's lifetime, Shakespeare had also written that wonderful line, 'There are more things in heaven and earth, Horatio, than are dreamt of in your philosophy' (*Hamlet – The Complete Works of Shakespeare*, New York, Doubleday, 1967 p. 606). As so often, good old William put his finger right on the button. What Galileo had spied through his new telescope was not just moons, but another Universe, a wholly different space to anything we had thought about before.

For philosophers, like everyone else, this meant that we had to question many of the things that we had taken for granted, and the certainty of the world seemed to be evaporating in front of our new lenses. What was happening was that, as more and more people looked at the stars with telescopes and discovered more and more weird things flying about, it became obvious that the Universe was neither simple, fixed, static or easily understood. The idea that the Universe was a set size, which never changed, had been extremely common since the first cavemen had drawn bad pictures on walls (was this graffiti?) and has a deep psychological appeal. Apart from the infamous Heraclitus, the early Greek philosopher who argued that everything was in flux, and thereby claimed that the sun and the planets would probably blow up in due course, almost all thinkers and astronomers assumed that the Universe had always existed and would continue to do so. This idea of a finite, static Universe underpins every conservative view of the world that has ever existed, from the Babylonians to the Texans. From a general idea of a fixed Universe, all sorts of reactionary ideas can be assumed, like the notion of fixed male and female identities, or of gods and kings who rule things by natural decree. This is an important philosophical point that the Greeks realised. Ideas of the static and unchanging reality of being and nature lead more or less directly to repressive views of fixed political systems (idealism breeds repression. as someone once said), which empower elitism and unthinking political control. Thus, there is a connection between the stars of the nightly firmament and the leaders of the current political elite, and science and politics interact precisely in thinking about the fluidity of all things, particularly who owns and controls what. The famous moon landing in 1969 was as much about proving America's technological superiority over Soviet Russia as it was about exploring space, and was America's revenge on Russia for putting the first satellite into space in the 1950s. The Chinese have just announced that they are getting into space as well, and this is real geopolitics of the universal kind.

The idea of Galileo throwing balls off the Leaning Tower of Pisa, or feathers or whatever, is a nice story but almost definitely not true. However, his work was to lead directly to Isaac Newton's famous work on gravity, which then got turned into another legend. What Galileo probably did was to roll different sized balls down a slope and then ascertain that balls of different weights moved at the same speed, or that some strange force acted on things equally. This was like thinking in miniature about the forces that made things whizz about in the whole Universe, like what made planets orbit, for example. Newton, who it is often said 'invented' gravity when an apple fell on his head (as though apples had never fallen on anyone else's head before), actually worked out the laws of motion of the Universe and theorised how the force of gravity might operate. It is possible that Newton was thinking about what caused apples to fall to the ground and that led to his theory of universal gravitation, but it was the mathematical mind that led him to the explanation, not a Cox's Pippin. (Weirdly, apples feature in a lot of legends from Adam and Eve onwards!)

Newton's interests were extraordinarily wide, from alchemy to religion, but his science and mathematics were straight-out genius level, from the invention of calculus to the laws of motion and the theory of gravitation. During one short period, from about 1665 through to 1690, Newton transformed mathematics and science, brought the Copernican revolution to completion and put forward the theory of gravity. Newton was a secretive sort, however, and everything had to be dragged out of him. He wrote huge amounts and published practically nothing in his life. One of the strangest and most important meetings in the history of cosmology was in 1684 when Edmund Halley turned up in Cambridge to have a chat with the great man about mutually interesting things and asked him about a difficult problem concerning the curve of planets and the sun. Newton more or less said, 'Oh, I solved that one ages ago,' but then couldn't find the proof, so he agreed to do it again and publish it. Halley kept hassling him and he eventually produced his mathematical principles of natural philosophy, which should have been called *Unbelievably Important Mathematical Principles of Everything* but which he called *Principia* and avoided publishing anything else. Before Newton, the question of understanding the Universe was still dominated by religious, philosophical and political considerations; you could still be burnt for questioning the notion that the sun and the planets went around the earth. By the time of Newton's death, it really was an established fact that the sun was the centre of the Universe, and the earth was just one little planet that ran along with the others.

The Newtonian Revolution:
Mechanics and Maestros

Never before, or since, has one person so completely rewritten the rules of the game, made scientific discoveries of world importance and fundamentally changed the way we view the Universe. Newton's mathematics led to the absolute recognition that the earth was just one minor planet in a vast Universe and that the forces that controlled the Universe had nothing whatsoever to do with humanity. He himself was basically a bit strange, easily distracted and obsessed with weird religions and turning lead into gold, but that demonstrates another law of the Universe – that being really normal is for bank managers and bean counters. With Newton, we truly enter the scientific age, one in which the conception of the Universe, which came to be generally accepted, was that of a giant self-regulating system in which the hand of God was no longer required. To blaspheme, we might say that gravity replaced God as the glue of the Universe, and that once we could think of it in non-human terms, we could imagine its immensity properly.

Newton's contribution to world science and mathematics was, unusually, recognised at the time and, with the publication of his *Principia Mathematica* in 1687, he was hailed everywhere as a genius. With the printing of this sexily titled book, a complete and powerful new theory called classical mechanics was born. In *Principia*, he outlined his laws of motion and his theory of gravitation, but because there wasn't an appropriate mathematics to explain his new laws, he was forced to invent the calculus, which he did, but kept it secret. (This led to an endless row with the German philosopher Leibniz, who claimed he'd thought of it first!) This trick of discovering things but not telling anybody was one of Newton's more endearing characteristics (and he had plenty of nasty ones), and some of his work was still being unearthed in the twentieth century (and some of it was very strange). It was only in 1936 that his wacky stuff on alchemy was unearthed by the great economist John Maynard Keynes, who found hundreds of pages of scribbling on how to turn lead into gold in an old trunk of Newton's papers. How, you might ask, could a genius like Newton believe in such mumbo-jumbo as alchemy? And if you can answer that, you've probably read Freud on the strangeness of creativity. Actually, in retrospect, alchemy isn't that weird because we now know that certain chemical forms can mutate into other things. It's just not lead into gold.

So what did Newton explain? Well, in general, he

put forward three laws of motion, which explained how, why and in what manner things moved about in the Universe, and then added the universal law of gravitation, which explains how everything in the Universe is attracted to everything else. It's the love-glue of molecules and goes right up to planet sizes, the 'whirling-force' that holds things in place. With space travel where you get weightless, or non-gravity situations, you can see what a fundamental aspect of everything gravity really is. Newton not only clearly explained how the planets orbited the sun in ellipses, but his universal law of gravitation showed precise calculations of how every object attracts every other object and that mass dictates attraction. Put vaguely scientifically, we can say that the force is proportional to the mass of the object and inversely proportional to the square of the distance between them (size and distance determine the gravitational pull). Despite being such an important force, as it were, gravity is relatively speaking rather a weak force, compared to something like a nuclear force, for example. This gravity thing was a universal law, Newton said, because it applied equally to small things falling on earth as well as to a planet orbiting the sun. This was clearly the first completely universal law in the history of mankind's knowledge of the Universe and is thus the absolute turning point in our world view. Newton left us with the idea that basic mathematical principles actually work, in all places and at all times, to explain the Universe.

Newton's work opened up the explanations of so many things scientifically (including the fact that the earth bulges a bit in the middle because it spins round), that he quite rightly is the man who had the force with him. That he explained what this force was in such simple and seemingly correct terms that it did apply to everything is really something of a miracle. And he was from Lincolnshire. The first law of motion said that a body moves in a uniform motion, so that if nothing else interferes with it, it will just keep moving forever in a straight line at the same speed. The second law of motion equally said that any force that acts upon an object will do so proportionally and in the direction that a force acts upon it. Or to put that simply, there are precise, measurable ways that forces act in the Universe. The third law of motion simply states that for every action, there is an equal and opposite reaction; again in other words, that there is absolute regularity in the way things work. These Newtonian mechanics, based on universal laws, seemed to many people to be the answer to everything and thus inaugurated the great scientific age in which numerous laws were discovered and science became, as they say, the new black. Newton did all of this whilst at Cambridge University, where he was a fellow, and did work on many things including optics, theology, alchemy and calculus; he also did a lot of study of obscure religions and maybe invented Nostradamus as well. He was absolutely the least likely scientific genius one could have dreamt up. He was, apparently, neither attracted to, nor attracted by, members of the opposite, or any, sex and when they made him head of the Royal Society, he proceeded to act like a miserable old bugger. There's probably another universal law there somewhere as well.

Newton was not only a genius (and if you don't know what that means, try reading *Principia Mathematica* straight through), but he was also relatively modest, famously saying about himself: 'I do not know what I may appear to the world; but to myself I seem to have been only like a boy, playing on the sea-shore, and diverting myself, in now and then finding a smoother pebble, or a prettier shell than ordinary, whilst the great ocean of truth lay all undiscovered before me.' Modern physicists don't always seem to be quite so humble (except Einstein, who was similarly self-depreciating) nor have any of them contributed so much, in so many fields. Newton's work on light really kicked off the science of optics and he also invented a new kind of telescope, the reflecting type, using a mirror, which was of great importance in sky-watching. Perhaps because he was a slight obsessive, or because he had a beautifully ordered mind, Newton turned all of the vague approximations of the past into elegant and very precise rules about how to measure things. Whereas the

Greeks might have said, 'Oh the moon is a good three weeks flying away,' Newton developed mathematical formulae that pinpointed the exact ratios of things, provided you measured them properly. After Newton, rough approximations were out and exact rigour was in, and it still seems odd that we can now measure galactic distance but we can't make the trains run on time. Newton also, of course, postulated that there was both absolute space and absolute time, unchanging and immutable. This is the scientific certainty he bequeathed us and, allied with endless technological development, it produced a sense of certainty about the Universe that led to a kind of mechanistic arrogance, which only Einstein would puncture.

The man who persuaded Newton to go public, Edmund Halley, also paid for the *Principia*'s printing, so there are at least three counts on which our view of the Universe is indebted to this energetic character. Actually, if you add his discovery of a comet, it should be four. Or possibly five if you include his work as Royal Astronomer. One of the major reasons why we are indebted to Halley is that he became curious about the position of the stars and started thinking about whether it had all moved since Ptolemy had compiled his catalogues. He studied Ptolemy's *Almagest* and then compared the stars in the heavens with what he could see, some of which he observed from the island of St Helena. In 1718, he reported that, in fact, several stars appeared to have changed their position relative to the earth. This observation implicitly suggested that things in the Universe moved around more than anyone imagined, and that something even stranger was going on than currently imagined. The notion of natural perfection, of a neatly Christian little Universe going around in perfect circles, was beginning to look about as realistic as Joan of Arc's dress sense. The fact that stars had changed their position over the centuries was puzzling, and actually implied that the Universe wasn't static. That, however, was a question that astronomers weren't ready yet to answer.

After and before Newton should be described like am/pm, or BC/AD, for the divide is so complete that the two are two different worlds. For the next 250 years, people just filled in the missing bits and used the model set up by Newton. We call this a paradigm, or the set of rules and arguments everybody thinks within, and it defines the assumptions and discussions that people in a particular era engage in. We can safely say that we lived in a Newtonian paradigm right up until Einstein's fun and games in the early 1900s. The other great things about Newton's laws of the Universe were that, since they applied to so many areas, there were endless things that could be measured and analysed, which is what many scientists like to do. Since this was also the beginning of the era of the development of technology that would lead to the Industrial

Revolution, it also meant that there tended to be new methods of measuring and analysing as well.

Everyone was busy measuring the distance to the moon, the sun, the stars and anything else in the Universe they could think of, including comets and planets that hadn't yet been discovered. Everything that was successfully measured seemed to prove Newton to be right, particularly his laws of motion. Once again, mathematics had proved to be the secret code that unlocked the power of the Universe, but we always had to look in the right place, at the right time and with the right instruments. As science became allied with technology to produce seemingly endless new knowledge, and wealth, people began to feel happier with this notion of a mechanical, and knowable, Universe. Descartes, the French philosopher and mathematician, had used the analogy of a clock in talking about the Universe and this became a fairly general way of looking at things. A wholly deterministic Universe that moved like clockwork suited everybody really. You could even put the Creator in as the person who wound it up and let it go. However, you can't make a Universe without breaking eggs, as Kepler might have said.

The Rise of Modern Cosmology:
From Here to Eternity

The popularity of science and studying the Universe led to a scramble for setting up royal observatories, which became something of a nationalistic star race, and the development of proper astronomical telescopes made studying the Universe a much more professional business by the time of Newton's death. Mind you, many in the Arab world had developed observatories much earlier, like the observatory built for the famous Persian astronomer, Nasir al-Din al-Tusi, in Iran in 1259. Or the famous Ulugh Beg's observatory at Samarkand, which produced astronomical tables that included a catalogue of over 1,000 stars in the 1420s. Somehow, the spread of Western empires during the sixteenth and seventeenth centuries led to a certain kind of eclipse of the East's contribution to knowledge and to the reinvention of the astronomical wheel. This new phase of building observatories in the West, however, was driven by the new scientific fervour that was sweeping across Europe, and was also a product of the new astronomical technology that made observation so much more accurate and all-encompassing. The race to build big telescopes was quite reminiscent of the space race of the sixties and seventies, but of course, it was pretty hopeless having them in Northern Europe where it was cloudy all the time. Eventually, a few people realised that observatories would be better placed in tropical climes where good weather was a much better possibility. While our notion of the Universe was rapidly expanding, it seemed as though our cultural horizons were diminishing. This was becoming the West and the rest.

The first official Western scientific institution was formed in 1657 in Florence, and was called the Accademia del Cimento. For some reason, it lasted only a decade before they all fell out. Then followed the Royal Society of London in 1660, which was given Royal patronage and became just the Royal Society. Not to be outdone, the French quickly set up the Royal Academy of Sciences in Paris in 1666, and soon everybody had one. It has to be said that one of the real motivations of all this was to improve sea navigation, and thereby rule the waves, as Britain was doing quite well in any case. To get exact navigation, you needed exact measurements for the stars and planets and to understand how the heavens moved, so there really was a very down-to-earth reason for looking at the stars. The recent entertaining best-seller, *Longitude: The True Story of a Lone Genius Who Solved the Greatest Scientific Problem of His Time* by Dava Sobel, pretty comprehensively covers this area. Scientific advance is always a mix of genius, politics, disaster, good luck and financial gain, as poor old Kepler found out the hard way. In the late eighteenth century though, science was the new black, and the Universe was the new white.

Naturally, the Church did not like science setting itself up as a major opposition, particularly as it claimed to explain the Universe in objective terms and thereby do away with God. So it fought back. God was the creator of the Universe, it was argued, and its complexity simply showed the greatness of God's mind. This was known as the Designer Argument, i.e. that God was not a naff designer. This is a perfectly reasonable argument, except for the fact that the Catholic Church, in particular, still said officially that the earth was the centre of the Universe. Newton's arguments seemed to suggest that the Universe worked by itself and didn't need a creator to keep it going, although he in fact believed that God had set the Universe up and did routine maintenance, a position that still holds sway in some quarters. However, many concluded that lifeless matter was moved by molecular force, and the hand of God wasn't very clear, even if you accepted the idea that he kicked the Universe off. The ghost of Galileo stalked the corridors of scientific societies and churches alike; God and the physicist were in opposite corners. The trouble with very good telescopes was that you could see where heaven and hell were supposed to be, which had never been possible before, and in fact, there were lots of stars and absolutely no angels. What we have here is an increasing recognition of the sublime geometry of the Universe, amazement at all new scientific discoveries, and a kind of scariness about where it all leaves us.

This battle between science and religion became officially known as the Enlightenment, which is a pretty self-explanatory idea, and is really a period in which scientific reason took on, and replaced, religion as the dominant mode of thought in Europe. This age of the Enlightenment was enormously affected by Isaac Newton's discovery of universal gravitation and by Galileo's discoveries. The general argument was that if humanity could unravel the laws of the Universe, through the application of reason, why could it not also unravel the laws that ruled nature and society as well. This was a kind of mechanical optimism that went: Newton + philosophy = social progress and knowledge. People like the French Encyclopaedists started to argue that through proper education, and the application of scientific knowledge, humanity itself could be altered and improved. This idea of the rule of reason seems a little quaint from where we sit now but you can see the appeal of a regulated, mechanical and scientific Universe in which everything fitted in its right place. It's also true that these social scientists were probably looking through the wrong end of the telescope when they conceived of man as being an entirely rational being. That other universal law, 'there's nowt so queer as folk,' springs to mind here when talking about scientific optimism. It is claimed by some that this new reason led to the French Revolution itself, in which a new vision of humanity was put forward, but there was rather a lot of old-fashioned nastiness and cutting off of heads as well, so the idea of a rationally planned society took a back seat for a while. It was left to the nineteenth century to try and invent a science of society, and to the twentieth to prove that it was an idea whose time would come when pigs mastered the law of gravity.

Back in the world of cosmology and science, most of the actual scientists attempted to ignore the political and social upheavals that were going on around them, just as Newton had done, and to concentrate on the many burning questions that the new cosmology threw up. Like working out the size of the earth, for example, or measuring the distance a star might be from earth, or identifying comets or finding new planets. Or trying to work out how far away a star is by analysing how much light it produced and then estimating that, if the star was like the sun, it might take blah number of years for the light to reach us. How large, or heavy, the earth was and the nature of longitude all became burning questions, and then someone raised the question of how old the earth was as well. During the eighteenth and nineteenth centuries, people traipsed all over the world to measure and observe all of these things, as wonderfully outlined in Bill Bryson's *A Short History of Nearly Everything*.

Basically, since Descartes and Newton, everyone had been looking at the Universe as relations of particles, matter, space and time and this was such an exciting and new way of thinking, that what used to be called the 'philosophy of nature' got thrown out of the window. It is also probably true that Protestantism and the new sciences tended to get along better than Catholicism and the new mechanical Universe. Philosophically speaking, ideas about the Universe got wackier all the time. Every new scientific discovery seemed to spawn a kind of mirror image anti-science theory that recast religion in strange new ways. It is hard now to show how strange science could appear to be in the eighteenth and nineteenth centuries, because today we accept certain 'hard' facts as real and know, mostly, which things are paranormal rather than just slightly normal. One quirky Englishman, Thomas Wright of Durham, published a wonderful work in 1750 called *An Original Theory or New Hypotheses of the Universe*, in which he put forward the strange notion that the Milky Way was a large block of stars which was controlled by a force of supernatural energy, containing a power of morality and wisdom. This is *Star Trek* for the eighteenth century and actually it was quite popular as well, supposedly influencing Immanuel Kant and

Herschel, the German astronomer who discovered Uranus in 1781. Thomas Wright thought that there might be lots of strange slabs of light out there, which were 'creations' of this kind that looked like faint clouds of light. This was seriously weird stuff for 1750 but of course he was absolutely right, so to speak.

Immanuel Kant was quite possibly the greatest philosopher of all time and so his view of the Universe is probably worth listening to, even though he was neither an astronomer nor a cosmologist. In 1755, he produced a work that dealt with the possible origins of the Universe, his *General Natural History and Theory of the Heavens*, which was clearly influenced by Wright's ideas. Kant argued that the Milky Way was a kind of optical effect because of our position in a slab of stars and that the other wispy bits were probably nebulae, or other galaxies outside of our own. He believed that the sun was formed originally from a whirling mass of gases and, as the temperature rose to millions of degrees, a star was born. Or to put that another way, through the process of thermonuclear hydrogen fusion, the sun began to shine. What is a little strange here is that the conclusion was bizarrely correct but there were no solid arguments or evidence to back it up in any way (at the time). Perhaps that is a reasonable definition of many philosophical approaches but, in any case, Kant combined it with a splendid set of arguments about the nature of life on other planets, which he concluded must exist and that their rational powers would be related to their distance to the centre of the Universe. The argument for this was that creatures nearer the centre of the Universe would be made of baser materials and thereby would be less rational than those on the further points of the Universe. The earth, Kant stoically explained, was almost exactly in the middle and we were therefore pretty mediocre, whereas people out there on Saturn were likely to be very bright. His approach to all this was certainly methodical, but later observations have not exactly backed him up on this theory. Rather like Kepler before him, Kant had quite a strong mystical sense of the harmonies of the Universe. He just did not approach them in a particularly scientific way. It was really a question of, 'Don't give up the philosophical day job'.

Herschel, on the other hand, while being similarly struck by the poetic discussions of Thomas Wright, wanted to throw light on the matter of the Universe by the now tried and tested method of looking through telescopes. He returned to the reflective telescopes developed by Newton, persuaded George III to make him the Royal Astronomer and built a 40-foot telescope (which actually turned out to be too unwieldy). More importantly, he divided the sky up into regions (700) and professionally mapped each region one by one, thereby putting cosmology for the first time on a proper scientific basis. What is particularly extraordinary about this is that Herschel was originally a professional musician, and explains why he held a concert inside his giant telescope to celebrate its opening night. Herschel was aided in these endeavours by his daughter, Caroline, to whom George III also granted a salary, probably making her the first female professional astronomer in history. The Herschels did so much in astronomy that it would take hours to enumerate it all. Among their star turns were identifying double stars, plotting the entire Universe, discovering moons on Saturn and Uranus (after discovering the original), discovering infrared radiation and cataloguing thousands of nebulae (star clusters). They also managed to keep putting on concerts as well. In recognition of their contribution, the Space Station Observatory was recently renamed the Herschel Observatory. Perhaps astronomers should set up 'last night of the telescope' concerts to make themselves more popular with the public.

Herschel's discovery of infrared radiation raised the question of what parts of the Universe we actually see and, indeed, the question of light itself was increasingly becoming an important issue. In fact, from the time of Euclid, there had been discussion of optics and Ptolemy himself had discussed the problem of the 'refraction' of light in the atmosphere. Kepler had speculated that the speed of light was infinite and put forward a theory of lenses that led to the astronomical telescope, and Newton did early important work on how white light splits into its component colours when it is passed through a prism. Understanding the nature of light was becoming imperative, especially in terms of understanding what we actually saw when we looked through telescopes, but progress was very slow during the eighteenth and nineteenth centuries. There was also the question of the sun, the most obvious source of light for our planet. In fact, our sun is just another star, but it is also an atomic furnace that turns mass into energy. (It is estimated that every second, it converts over 657 million tons of hydrogen into 653 tons of helium.) This truly spectacular activity leaves a missing four million tons of mass, which are fortunately discharged into the galaxy as energy (again, fortunately the earth receives only about one two-billionths of this). It is not surprising that the Egyptians and Mayans worshipped the sun as a god; we certainly couldn't get by without it. Perhaps it is more surprising that many more cultures didn't plump for the sun as their main god, although all tourists now seem to have reverted to sun-worship. Currently, it is estimated that the sun should keep burning for another ten to thirty billion years, which is the good news. The bad news is that with the advent of global warming, we'll all have fried by then and tourists won't have to go to Spain. They'll be able to sunbathe in Grimsby. You will remember that Kant had theorised that the Universe was all dark to start with and then, after the formation of stars, the gaseous nebula, there had been light, generated by the heat of energy transfer. But the question kept nagging, what exactly is light? As Homer Simpson said, 'You turn on the switch, and then you can turn it off again.'

Euclid in his *Optica*, written about 300 BC, had argued that light travelled in straight lines and had thought about reflection, which really became an issue again with telescopes. Roger Bacon, in the thirteenth century, put forward a theory that light travelled like sound, which was an interesting speculation and one that was around at the same time spectacles were mentioned. So the technology and ideas around telescopes were actually knocking around for a couple of hundred years before anyone thought to put them together. Many great minds thought about optics and reflection but the next most important date in the history of our understanding of the Universe is 1676, when one Ole Christensen Roemer, a Danish astronomer, thought of measuring the speed of light. Naturally, Newton had begun the science of optics and, in his attempts to analyse the light from stars and compare them to the sun's light, he prefigured much later work. Roemer concluded that it travelled at a finite speed, which he estimated at 140,000 miles per second. Absolutely no one at the time took any notice at all, but measuring the speed of light was to become of huge importance, particularly to Einstein. Roemer was studying Jupiter's moons at the Royal Observatory in Paris and he noticed that at different times the moons seemed to vary the length of time they took to go around. Being a good scientist, he realised that these times varied with the distance from earth so the possible explanation was the length of time the image took to reach earth, or in other words, the speed of light and the distance. Ole then tried to work out the speed of light but unfortunately his distance calculator to Jupiter wasn't the best, so he came up with a speed of light of 140,000 miles per second. Although he was out by 46,000 miles per second, it was a brave attempt, but the implications of what he had discovered were not to be realised for a long time. Thinking about the stars, and the light they emitted, was to become one of the key areas of later astronomy and so theories of the nature of light were like a new code that had to be cracked.

Much later, thinking about where light and radiation came from was also to lead to the discovery of the idea of the birth of the Universe, via a few more complicated developments and theories. But light, that strange and varied thing, eventually turns out to be the only constant thing in the Universe. In thinking about light, there is also what is known as Olber's paradox, which is the question of why it is dark at night. This is really a question about how light moves around the Universe and how, if all stars were similarly bright, you might think that the sky would be permanently bright. This was thought to be a problem because, if you accepted that the Universe was static and fixed, the light from all of those stars would fill the sky, even at night. Various wacky proposals were put forward. Perhaps solar dust absorbed all the light or some of the stars were just really dim. Kepler, Halley, Lord Kelvin and even Friedrich Engels, the great communist writer, had a view on this paradox and it was really only the Big Bang theory that blew the whole question out of the water (or the sky?). The point was that the Universe is not static, and stars, light, galaxies and other things are in constant states of development and, as we now know, the light we see comes from all over the place. Everything in the Universe, it seemed, was getting more complicated by the minute and in reality, it had all yet to kick off.

Someone else who threw light on the nature of the Universe, but at the other end of the scale, was good old Marie Curie, non-cosmologist and discoverer of radiation, which she called radioactivity. She set out with a simple research hypothesis, that radioactivity was a property of atomic structure, and brought about a fundamental shift in scientific understanding, for which she won two Nobel prizes. At the time she got started, the 1890s, almost all scientists regarded the atom as the most elementary particle in existence and, rather like the Universe, fixed and unchanging. It is always difficult looking back to see how much certain ideas are just accepted and thereby produce strange and complicated ways of trying to explain the consequences of the idea, rather than dumping it, like the Ptolemaic explanation of the Universe itself. Marie Curie cut through all of the messy and complicated ideas surrounding waves, radiation, particles and elements and established a new basis for thinking about atomic structure. This came about because there were several curious problems in existence when Marie Curie was looking for a research project. One of them was radioactivity itself, which nobody had a name for, or explanation of, but which seemed to infect any laboratory that worked with elements like radium or thorium. A German physicist called Wilhelm Roentgen discovered a strange ray that would pass through people and produce an image of their bones. He called them x-rays, and their use was immediately apparent. The next year, Henri Becquerel discovered uranium rays and called them, funnily enough, Becquerel rays, but it was left to the Polish postgraduate, working in a cupboard, to make the all important breakthrough about their real nature.

She was working in what had been a broom cupboard and, using a machine, an electrometer, invented by her brothers, she showed that uranium and thorium emitted 'Becquerel' rays. Marie went much further, however, and formed a crucial hypothesis: that the emission of rays by uranium compounds was probably something to do with the structure of the uranium atom, which meant that atomic structure was likely to be more complicated than anyone had previously thought. This very neatly, in 1900, began the process of attempting to understand the structure of the atom, of radiation, of particle physics and of the atomic structure of the Universe. One more particle in the puzzle was being filled in but, as ever, it opened the door to bigger and more mysterious patterns in the Universe. Sometime later, Marie Curie got together with Einstein and they went sailing together on some Swiss lake. We don't know what they talked about but we do know that they got lost and argued about who was the worst sailor. Clearly, unlocking the secrets of the Universe doesn't always help you navigate. Now we have global satellite tracking systems that can tell you exactly where you are (in a traffic jam).

At more or less the same time as Curie, Max Planck was also doing his bit to complicate our picture of the Universe, by inventing something called 'quantum theory'. Unfortunately, Max Planck probably got overshadowed by Einstein, but he still managed to stuff up the entire Newtonian thing about matter and particles being deterministic and mechanical. To start with, energy in this new theory is not like those little diagrams of neat little waves buzzing about in straight lines. It operates in 'quanta' or packets, and it is also unpredictable. This tied up with the light thing because there was a realisation that it wasn't just a wave. It was beginning to seem like everything could be like something else. Waves could be particles or vice versa, like electromagnetism. Planck did some stuff on what was called the black-body spectrum and showed that indeed light could behave as if it were discreet packets, which acted like particles, but Planck just wanted to call them 'quanta'. Einstein picked it up and ran with it, saying that in fact light was made of particles, but could act like waves.

So when is a wave a particle? The answer may be when it feels like it, or when we look at it in a particular way. There is a thing called the Heisenberg Uncertainty Principle, which just about sums this area up, and slightly adds difficulty to the whole question of talking about light, matter, stars and distances. Actually, looking at stars was relatively easy. At least they were great big shiny things that more or less stayed in place and now you could point your telescope at them and take photographs as well, so you had something fairly tangible to consider. So easy questions like, 'How big is the Milky Way?' and 'Does it extend to infinity?' were simpler to deal with than all of this stuff about waves, particles, radiation and atomic structure. Someone called Schrödinger, who is famous for inventing a cat, also developed quantum mechanics in the twentieth century, which produced the need to have theories that connected up large-scale theories of the Universe, like Einstein's, and the small-scale stuff. This is sometimes called a GUF (Grand Unified Theory) and it's something we'll come back to later.

New Dimensions

In the history of actually discovering what we really knew about the Universe, the question of clearing up what that simple thing, light, was, looms large. This is basically because it is through interpreting the light we get from the rest of the Universe that we find out what is going on there. Newton had tried to estimate the distances of stars by their brightness and was correct in the assumption, but completely wrong in the method of applying it. The man who provided a (nearly) proper theory of light was James Clerk Maxwell, but this wasn't until 1865, and again the implications of his work weren't generally realised for some time. What this British physicist did was basically to unify the partial theories of electricity and magnetism that had been used to describe what went on in the atmosphere. This was a recognition that wave-like patterns played a key role in the movement of energy, a kind of underlying structure. He did this (mathematically) by showing that the two forces have a more or less common origin and, in fact, since that time, we speak of electromagnetism.

In his theory, this means that the emission of radiation by matter must be as a result of the acceleration of electrical charges, moving in wave-like patterns, like the ripples in a pond. His theory also predicted that light waves should travel at a fixed speed, and that would be relative to the substance called 'ether', which everyone still agreed existed. To put this crudely, it was simply a way that everyone operated to allow them to think of a stable Universe, but 'ether' was a sort of airy, non-reactive substance that allowed light or whatever to move easily. The fact that it didn't exist was a sort of theoretical problem; thinking of light and other wave patterns properly did away with the need to believe in this weird building-block. Maxwell also pointed to the existence of other regular wave-like patterns, what we now call radio waves, x-rays, gamma rays and so on. From his equations, it seemed that within matter itself there were mobile electrical charges that, as they moved in some way we didn't know, produced the spectrum. Marie Curie's later work on radioactivity would also open all of this up to science, but Maxwell's was the true pioneering bit on light.

As, seemingly, with all scientific theories, this was a great advance on previous arguments, but it immediately produced problems as well. One step forward, three waves back. From Maxwell's work it appeared that, if the fixed nature of the 'ether' were correct, then the speed of light would turn out to be different when measured as the earth was moving towards the source of the light. Those who worried about these things just assumed that, like many other things, proper observation would show that, indeed, light moved in different ways. Two real spoilsports in the shape of Michelson and Morley, one of whom won a Nobel prize for physics, measured light exactly in these different conditions and annoyingly found that it was always the same. This shocking result came in 1887 and just as we thought we had it, bugger, someone goes and tips several hundred years' worth of study down the drain in one little set of experiments. For many years after this, the best brains in the business tried to make the theory about light and the observations fit together. But could they? You'd have to be an Einstein to figure that one out.

Fortunately, Einstein had been born and was now working in the Swiss Army Knife patent office. Less fortunately, the problem about the nature of light turned out to be just one question in an inter-connected whole to do with matter, energy, acceleration, particles and eventually electrons, photons and neutrons etc. The Universe was about to go pear-shaped on us and, to misquote the poem, it wasn't waving, it was being a particle.

Thinking in general about the nature of the Universe at the beginning of the twentieth century was a bit like thinking about the future of the motor car. It was a newfangled thing that existed in a world that everyone thought was still somehow fixed. Some people saw that the world was whizzing along at a furious rate and others saw horses plodding along as they always had done. It all depended where you stood. That was where Einstein came in. He suddenly made everyone realise that once again we had been looking at things in a predetermined and inelastic way. Newton's fixed principles had worked for what seemed like forever, but in fact the problems had accumulated and it was Einstein's job to light the fuse-paper that would blow the whole thing apart.

It seems strange that, in 1900, some scientists were more or less arguing that most things had been discovered and that the mysteries of the Universe were just a matter of a bit more collecting, observing and cataloguing. Indeed, it does seem as though there is a terribly powerful human impulse to want the world to be knowable and controllable. Yet every time it starts to feel secure, somebody turns the world upside down. Einstein did this once again by bringing together many strands of thought and knowledge and drawing the obvious conclusion, just as Newton and Copernicus had done hundreds of years before. The only trouble this time was that Einstein pushed the world into a scientific complexity from which there was no escape, otherwise known as the Special Theory of Relativity. The real trouble was that the obvious conclusion was very difficult to develop, even harder to understand and led to forms of knowledge that were not only outside the box, but proved that the box was actually a sieve.

The great joke about Einstein was that when he was a kid, somebody (a bureaucrat or an educational psychologist, undoubtedly) told his poor parents that he was probably a bit retarded as he didn't talk (much) until he was three. Then, when he left university, the only job he could get was in the Swiss patent office where he beavered away, thinking about physics and filing the forms for the invention of Emmental cheese and cuckoo clocks. From these inauspicious beginnings, young Albert crafted a career as the twentieth century's best theorist of everything in the Universe, and the world's first mega-celebrity scientist – not that he particularly wanted or liked the fame. In 1905, the same year as the practice Russian revolution that would eventually turn into the real one, Einstein submitted some scientific papers he had knocked up in his lunchtime to the *Annalen der Physik*. This snazzily titled magazine fortunately recognised a good thing when they saw it and published his work, particularly the paper entitled *On the Electrodynamics of Moving Bodies*, which blithely rewrote Newtonian mechanics. Einstein had obviously been thinking very hard indeed about these problems, as the astounding level of thought in the paper demonstrated, and the extraordinary conclusions showed. Later in life, Einstein claimed that his main working method was a pencil and a piece of paper, especially for the famous equation $e=mc^2$. (This is something that computer-dependent junkies could think about.) Science, as we know, proceeds in paradigm shifts when all the rules are changed, and this was the definitive paradigm shift, even if it took a while for it to be recognised. Despite the fact that it was claimed that only three people in the Universe understood Einstein's relativity theories, by the 1930s, half of the Universe knew about them and he was recognised everywhere.

What, then, did Einstein have to say about the Universe? Well, put colloquially, what he was saying was that space and time are not absolutes, as most people had presupposed at least since Newton, but relative. How on earth can time be relative, you may well ask? There are sixty seconds in one minute, sixty minutes in one hour and so on, right up to one hundred years in a century. Well, indeed, that does seem to be the case but again, it is all about how you measure it. One of the most famous examples used to illustrate this was that of trains. If you've ever stood on a platform and watched an express whistle past, you'll have noticed that the sound arrives, changes and then lingers, and if you imagine the people on the train looking at you, you'd seem to be moving very fast as they flicked past. If the train was travelling at, say, 93,000 miles a second, or half the speed of light, the passengers and you might notice the relative difference of things and time would seem quite different depending on whether you were on the train or not. If you were on the train and you didn't stick your head out the window, you wouldn't really notice anything, but your time-space reality would be different to that of the guy on the platform who would seem to be acting in slow motion. You might argue that the guy on the platform is moving and you are stationary.

Einstein also showed that time and space were related and could be fickle, depending on the observer. In other words, if a chicken crosses the road at high enough speed, it might seem to someone on a train travelling at even greater speed that the chicken stayed still and the road moved. Space is like the glue in which time sloshes about and sometimes it sticks in different ways; it gets bendy like when you're really, really drunk. That is not a very scientific analogy but you sort of get the picture of the trickiness of what we are dealing with here. Einstein himself supposedly liked a really bad joke about two old tailors who were talking about this new relativity thing (condensed version):

The one old tailor says, 'So, what is this relativity thing?' And the other guy says, 'Well, I heard it's like if you're sitting in the front room working away and young Hettie comes and sits on your lap and tousles your hair for five hours, it feels like seconds, but if your cranky, fat old mother-in-law comes and sits on your lap for ten seconds, it feels like forever.' The other guy says, 'And he gets paid for this, already, and wins prizes?'

A version of this is quoted in a fine biography of the man, which explains the science as well as can be expected for the general reader. Indeed, there are many lives of Einstein but the recent biography by Dennis Brian is rather more judicious than most.

So, what else travels at the speed of light? Er, light. So electromagnetic waves are light and light is an electromagnetic wave. It's just waves and particles, swings and roundabouts. Poor old physics has never been quite the same since, and neither has our view of the Universe, partly because one form of 'light' is also radio waves. The whole electromagnetic spectrum plays tricks with our three-dimensional view of reality, and art. Basically, Maxwell's equations led directly to the discovery of radio waves, and then the invention of radio, radar and television, and all the joys of electronic culture – the new Universe as it's known.

The Cosmological Considerations on the General Theory of Relativity paper, which Einstein wrote in 1917, the year of the actual Russian Revolution, set the seal on the Einstein revolution. The Universe was once again not quite what we had thought it was, and gravity turned out to be both everywhere and sort of nowhere, in the sense that it is really just a distortion of space and time. Gravity, you will remember, is the weakest force in the Universe, but also the most universal. Space-time relativity also implied that gravity was more of an outcome of the warping of this new hybrid rather than a completely fixed force. At the same time, we have to remember that Einstein was not really a cosmologist and he accepted the then generally held view that the Universe was stable, or in other words, that it wasn't expanding or contracting. He used something he called the 'cosmological constant' in his work, a part of his theory that he later admitted was a big blunder. This mathematical formula expressed a more general notion, still generally held, that the Universe was in a kind of steady state, an idea that goes right back to the earliest human beliefs and the Aristotelian view of a perfect and complete Universe. It is a measure of how difficult thinking about the Universe had become when Einstein wasn't quite sure what was going on.

After Einstein had altered our sense of reality by asserting that space-time was the same thing, the twentieth century got to serious work on making the Universe more problematic, by going further out into space and deeper into the peculiar nature of matter. Actually, after Einstein, it was all a case of hubble, bubble, space and trouble, but mainly Hubble. If self-importance were a feature of the Universe, then Edwin Hubble would have had a galaxy all to himself and, considering that he discovered many galaxies, this would probably be fair enough. Hubble was one of those people who was incredibly good at everything. Rich, handsome and successful, he was the sort of person we all love to hate on the grounds that no one should be that good and know it. Irrespective of that, in 1929, he observed that distant galaxies of stars seem to be moving away from our galaxy, as well as away from each other, just as if the entire Universe were expanding (this became known as Hubble's Law). These observations at the Mount Wilson Observatory were to spark another revolution in our sense of the Universe and lead eventually to the Big Bang theory. The pipe-smoking, opinionated, athletic, English-loving, tall-story-telling Edwin Hubble somehow managed to frame incredibly important questions about the Universe while annoying everybody. That he pinched some of the information and ideas from other people goes without saying but he did ask, 'Hey, how old is this goddam galaxy, and just how big is it, and how many of them are there out there?'

The story starts with the wonderfully named Henrietta Swann Leavitt, and goes on in that wonderfully depressing way that these things do in our Universe. In the early years of the century, Henrietta started working at Harvard's Observatory for 30 cents an hour. She was paid to observe stars and compute where they were. Being smart, Henrietta noticed that certain stars changed their brightness from time to time, which would be odd if they were in a fixed place, space and time relative to us. She called them cepheid variables as they were stars that went through cycles of brightness and darkness. She found that when observing a cepheid variable, she could relate the length of the brightness cycle to the size of the star but, naturally, this could only be done with the most incredibly precise measurement. With this discovery, she was able in 1912 to work out the distances between stars and the earth by analysing the actual versus the changing brightness. This put stars at incredible distances from the earth and provided a method for comparing other stars and galaxies, a revolutionary breakthrough that Hubble grabbed with open arms.

Henrietta was ignored and died of cancer in 1923, whereas Hubble went on to make himself the most famous astronomer of the twentieth century. Using the new 100-inch telescope at Mount Wilson, which was up and running in 1918, he began chasing nebulae. He was trying to work out if these clusters, or patches, were made up of stars as several other people had claimed earlier. It was known that some of these spiral nebulae, little fuzzy patches of light in the night sky, contained individual stars, but no one agreed as to whether they were little clusters on the edge of the Universe or something else. In 1924, Hubble measured the distance to the Andromeda nebula, a fuzzy little patch of light that appeared to be the same size as the moon, and he blithely demonstrated that it was about 100,000 times as far away as the nearest stars. The only logical conclusion was that it had to be a wholly separate galaxy, and rather than being the size of the moon, it was comparable in size to our own Milky Way, but very much further away. This really was exciting stuff, and despite Hubble's probably outrageous luck in being in the right place at the right time, he certainly stirred up our ideas of the good old Milky Way. To put that into focus and to explain what Hubble did, you have to remember that the only galaxy we knew about properly at the time was our own, and we kind of assumed that that was it. Despite all the science, all the technology and all of the observing, there we were in the 1920s blithely doing the Charleston and assuming that our little galaxy was the Universe, with a bit of space at the edges maybe. It is now thought that there are more than 140 billion galaxies in the neighbourhood, or the ones that we know about, and good old Edwin pointed this out. Basically, in the early 1920s, Hubble played a central role in establishing just what galaxies are. To be fair to him, he was the founder of observational cosmology, he set up the system of measuring extragalactic distances and he proved that the Universe was expanding, in all directions. He announced what he called Hubble's Law in 1929, which said that galaxies appear to be moving away from us in all directions, and, what is more, the further away a galaxy is, the faster it seems to be moving. The lesson of this seems to be that if you have an attachment to a particular far-flung galaxy, you'd better go and visit it soon or it may be gone.

How did Hubble establish that these galaxies were moving away? Well, the short answer is by looking; the longer answer is by using the best telescope in the world and by using the newly established art of photographing the Universe and studying the pictures. In fact, he was only able to measure the distances to a few of the other galaxies, but he had worked out, courtesy of Henrietta Leavitt, that he could take their apparent brightness as an indicator of their distance. He used something called the Doppler shift to measure the speed with which a galaxy was moving away, or towards us, using a spectrograph which measures the tiny red shift of light. Christian Doppler had discovered in 1842 that sound waves had a different pitch depending on whether the source was moving towards you or away from you, like the sound of a train horn as it belts along. Hubble extended this idea to light, whose waves, he realised, worked in the same way. He found that the light from distant galaxies was shifted towards the higher frequencies, or the red end of the spectrum, which meant that they were moving away. The observational data available to Hubble in 1929 as he developed his ideas were patchy, in that only a few galaxies had been looked at, but either by genius or by extraordinary good luck, he immediately worked out that there was a straight line fit between the data points, which showed that the red shift was proportional to the distance. This is like walking into someone's twenty-year research programme and saying, after five minutes, 'Here's how it works, really,' and then walking out again. Almost without trying, Hubble solved the entire problem of how to measure the Universe, and showed it was vastly bigger than anyone had ever imagined. It's not known whether he said, 'Oh, I must invent modern cosmology before I go off and play tennis,' but he might well have done.

What Hubble left us with were the questions of why the Universe was expanding and what the implications of this were, and just what were these extragalactic nebulae? There were also the questions of how old the Universe was and quite importantly, what was the total mass of the Universe? Hubble's claim in 1929 that the further a galaxy was from us, the faster it was moving away, was the small bang that led to the Big Bang as it conclusively showed that the Universe was expanding. The new cosmology was getting trickier by the day, and even worse by night.

Holes, Bangs and Curvature: Eternity Gets Bigger

All of the early twentieth-century developments in cosmology proved only one thing: we universally knew less about the Universe than we thought. 1929 was a very bad year in many respects for the stability of the Universe. Hubble's Law clearly showed that everything was flying apart faster than anyone could have imagined and bubble's law, that what goes up goes down, showed that the Wall Street stock market only expanded until someone showed the holes in it. The Wall Street crash was the first known incidence of world markets imploding together, and of an economic recession of a severity unheard of in our galaxy. The news that our galaxy was only one little one in a rapidly expanding Universe may not have helped the mental stability of stockbrokers. I am not claiming here a connection between the Universe and the stock market, but there may be one. Hubble's observations, made with the help of a janitor assistant who sat out in the freezing observatory for months, showed conclusively that galaxies were moving away from us and away from each other.

This actually tied in with mathematical work that people were doing with Einstein's equations that suggested the Universe must be either contracting or expanding; it couldn't be sitting still (see *A Brief History of Time* for a slightly fuller explanation).

The pace of things was getting too much for many people and once again, philosophical reaction to the idea of an ever-changing Universe was strong. Oddly enough, Einstein himself had resisted the idea that the Universe was expanding, for reasons connected with his idea of a 'cosmological constant', but he eventually concurred in the face of the evidence, as all good scientists do. Living in an expanding Universe generally ought to bring down house prices, as there is always more room than when you started, but we do like to huddle together. As we could see more of the Universe with giant telescopes, it started to become apparent that we could not see the edge of the Universe but, perversely, we could see its past. Measuring the speed of light, it turned out, was also measuring the past so that in one sense, time travel is true. People were also beginning to think realistically about space travel as well, and this, of course, was where the politics started to come in.

During the 1930s, science became involved in politics in ways that hadn't really happened before, and even astronomy became political. Science fiction was scaring everyone, the Russian Revolution was particularly scaring capitalist America, and strange forms of pseudo-science, like eugenics, were being picked up by scary nutters like the Nazis. At the same time, jolly old Joseph Stalin loved Soviet science but he locked up one unfortunate, Nikolai Kozyrev, a physicist, for having views about an expanding Universe that didn't agree with the communist line. When Kozyrev appealed against this lunacy, they changed his sentence to death, this being one of Stalin's little jokes (they eventually let him off). The fact that several scientists had already speculated about using the new particle theory to split the atom and thereby also make an atomic bomb was also occupying people's minds in the late 1930s, an era that WH Auden described as, 'that low, dishonest, decade'. In September 1939, an obscure magazine called the *Physical Review* published one article that talked about the possible gravitational collapse of stars into themselves, and another article that talked about how nuclear fission would work, or how to make an atom bomb. That about summed up where the world was going and suggested a theory about the Universe which wouldn't be properly clarified for decades. Einstein thought this stuff about imploding stars was just science fiction, but perhaps he was just getting on, relatively speaking.

The Second World War wasn't very good for international scientific cooperation and cosmological research tended to get put on the back-burner, along with science, humanity and reason. But the Universe just kept expanding. In 1942, a bunch of people in the Manhattan project, based at Los Alamos, pulled off the first nuclear chain reaction, thereby providing the greatest demonstration ever of theory being put into practice. The work of Maxwell, Marie Curie, Planck, Niels Bohr and Einstein came together in unleashing the unbelievable energy residing in small-scale matter, in sub-atomic structure, in splitting the atom apart and showing that theoretical physics was far more than thinking in abstract terms about mathematical relations and structures. Nuclear reactions led to nuclear bombs and suddenly we had a whole new way of destroying ourselves, and if we wanted to, the planet. As someone once observed, it is very peculiar that at the same time as we discovered the endless immensity of the Universe, we also began to work out that the smallest known fact of the Universe, the atom, was actually neither the smallest nor the most obvious thing in existence. We have been happily splitting the atom ever since 1942, either in atomic explosions, nuclear reactors or in particle accelerators that blow atoms apart to see what happens.

There were inklings that knowledge about atomic structure would assist us in knowing about the wider Universe as well, and that is where the connection between the very small and the unbelievably enormous comes into play. This is very useful when you are trying to think about the origins of the Universe, otherwise known as the Big Bang, a sort of nuclear explosion so absurdly large it was enough to start a Universe. We hadn't quite got to the point at which a Big Bang theory was accepted by everyone, or even properly demonstrated but, if you knew where to look, all of the clues were pointing in the same direction. There was still opposition to the idea of the Universe having an actual beginning, rather than always having been there, and this was known as the 'steady state theory', put forward by Fred Hoyle, Herman Bondi and Thomas Gold (afterwards the BGH line), in which the Universe is in a steady state. Newton would have approved of this line, as indeed would have Aristotle but, from the time they proposed this theory in 1948 until the 1960s, it was treated seriously, only to be shown to be quite wrong by the discovery of a background cosmic radiation that permeated the entire Universe and which must have come from a big bang. The steady-staters were claiming that the Universe looked the same whichever way you turned and that it had always existed and always would. In order to explain the dynamism of the Universe, they argued that bits of matter were continually being formed to fill up the spaces and perhaps new galaxies formed to fill in the gaps. This had some appeal but it didn't really explain why the Universe was expanding quite so fast. The step from realising that the Universe was definitely expanding in all directions, as Hubble and others had demonstrated, to thinking back to why it was expanding, is not really that giant a step (more like a hop) but then the idea of an original bang, however big, is a fairly strange notion. In terms of twentieth-century developments of theories of the Universe, it may well be the case that it very much looked as though whenever a peculiar explanation of things came to light, it was likely to prove correct in the end. In the nineteenth century, scientists may well have said that the simplest explanation of things was the best; in the twentieth century this was reversed and, in the twenty-first century, it is almost axiomatic that nothing is what it appears to be.

The one thing that came out of the Second World War that provided a key in the development of understanding the Universe was radar. It could look even further than 100-inch telescopes and, as we got to understand radiation and its different forms, we could interpret more of what was going on out there. Fred Hoyle was instrumental in the development of radar and was famous in the fifties and sixties for his science fiction and his popularising of astronomical thought. He didn't like the Big Bang idea but was pretty keen on the notion of space travel. By the 1950s, the idea of getting into space physically as well was beginning to be taken seriously, and again the development of rockets in the Second World War made this feasible.

Rocketry actually goes back to Russia in the nineteenth century, when one Konstantin Tsiolkovskii wrote theoretical papers about how to make them work and how to have multistage rockets using liquid hydrogen and nitrogen to propel them, but it was the Cold War and the space race that made it all a reality. It was the Russians who kicked it off in 1957 when they sent Sputnik into space, a little sphere orbiting the earth, which struck terror into the hearts of the Americans. This little round tin can, about two feet across, carrying radio equipment and weighing about 184 pounds, passed over America and the government had to reassure its citizens that it didn't carry nuclear bombs or anthrax or something. The second Sputnik, in November 1957, sent up a dog, Laika, and presumably the Americans were worried that she was a spy, but as far as we know, she didn't reveal any secrets on her return to earth. The Americans pulled out all of the stops and got a satellite into orbit in January 1958, which carried out some scientific analysis of the stratosphere, but the Russians upped the ante by putting a man into space in 1961. Yuri Gagarin is the most famous astronaut of all time and was a hero of the Soviet Union – the first human in history to enter space. Despite the Cold War lunacy of the space race, a great deal of information was rapidly produced by satellites and, by 1966, we had non-manned landings on the moon, followed in 1969 by two Americans landing on the moon (there are those, of course, who believe that this was faked). In 1970, the Russian Venera 7 soft-landed on the planet Venus and true deep space exploration was underway, reaching a peak with the launching of the Hubble Space Telescope in 1990 and the recent orbiting space station, run through international cooperation. All of this exploration has produced massive amounts of information about the Universe, but at enormous costs, and there are those who oppose space exploration, like Bertrand Russell, the famous philosopher. He argued that it was a waste of resources, driven by the military and the Cold War, and didn't improve human understanding. He may be wrong in this case, however, because in 1992, the Pope announced that actually Galileo's work was correct and that the earth does go around the sun, so there is progress.

To bring the discussion back to earth, it's necessary to think back to the theoretical basis on which thinking about space and the Universe took place. The arguments about what the Universe was doing go back to Einstein and general relativity, and it is interesting how often things do come back to him. A good number of the remaining questions are related to the mathematical problems that are thrown up by his field equations. These set out to explain curved space and the distribution of mass in the Universe, where all that stuff comes from and what happens to it. These equations are a bit tricky and, as it turned out, there are different answers to them, which even puzzled Einstein. (Why invent equations you can't solve?) Fortunately, another Russian was at hand to solve the tricky bits. Alexander Friedmann was a young mathematician who fought in the First World War and then conducted his research during the Russian Revolution. In 1922, Friedmann showed that, by using Einstein's equations, the Universe could either be shrinking or expanding, but what wasn't an option was Einstein's notion of a static Universe. You will remember that Einstein had talked about a 'cosmological constant' and assumed that the Universe was in a kind of steady state, and old Albert reportedly said of these conclusions about an expanding Universe, 'to admit such a possibility seems senseless'. It's a good job Einstein put that 'seems' in there because as we know, what seems strange may well be more accurate than things that seem sensible.

Friedmann actually found several answers to the question of general relativity, each one outlining a different possible Universe; not good news for the supporters of a simple theory. What Friedmann did, which was also so elegant, was to say the point is that we should assume that the Universe is the same all over the place and looks the same in every direction. This made it easier to think about the space-time problems and also implies that the nature of matter is the same all over the Universe. This makes it easier, apparently, to think about open and closed Universes in the field equations, to get rid of the idea of a flat Universe and to leave only the expanding Universe possibility. Friedmann died at the age of 37, before he could do much more, but he had ensured that Einstein had something to work on in his old age, like whether his own equations really did predict an expanding Universe. Then Abbé Lemaître (see below) came up with similar conclusions to Friedmann with the field equations but he was much more prepared to make predictions about what it meant. What this means, he said, is that there was a kind of cosmic egg from which it all started, and pouf!, there was a big bang and the Universe came into being. It's quite simple really, apart from identifying when this big bang happened, and also that rather tricky question of exactly why, and how. The latter are philosophers' questions. A theoretical physicist might just say that it's a fact that the Universe started with a big bang and we can assemble the evidence to prove it. The final proof came in the 1960s when Friedmann's thesis about the sameness of the Universe was demonstrated in the most wonderful way. Einstein had once said, 'God does not play dice with the Universe,' presumably implying that it was all quite well organised, but someone later added that God might well have lost the dice altogether, or at least misplaced them.

The debate about whether God had lost, forgotten, or never had the dice is incidentally taken up in Douglas Adams' great book, *The Hitchhiker's Guide to the Galaxy*, which is not only very funny, but also catches brilliantly the strangeness of our Universe, and the peculiarities of our place in it. He encapsulates how our scientific knowledge, which has accumulated over the centuries, somehow completely surpasses our individual ability to make sense of the galaxy as we live in it on a daily basis. The discovery of cosmic background radiation in 1965 could actually have been written by Adams himself, and demonstrates the curious way that knowledge inches forward in the real world. Back in the early part of the century, a strange character called Abbé Georges Lemaître, a Belgian cosmologist and Roman Catholic priest, did some work on Einstein's equations. He decided that, in fact, the Universe was expanding and that it must have started off at some point the size of a peanut, well actually a largish star, and then expanded massively. It's not known whether he checked these ideas with the Pope, but he certainly got laughed at when he announced these ideas, despite their theoretical basis and Friedmann's proofs of the expanding Universe. Abbé Lemaître had effectively invented the idea of the 'Big Bang' and he also suggested that, if he were right, there would be radiation left over from the initial Big Bang that would be traceable in the Universe. Lemaître apparently got hold of Einstein and Hubble in 1931 and explained his ideas at a seminar, which impressed Einstein while good old Hubble was already busy finding evidence that indeed the Universe was expanding.

The idea that the Universe started with a big bang didn't exactly get good publicity and most people ignored it, more or less politely, but a few people thought about the background cosmic radiation stuff. In fact, its existence was first predicted by George Gamow in 1948. What happened next was that two Bell laboratory researchers, Arno Penzias and Robert Wilson, were using a giant communications antenna at Holmdel, New Jersey, and they couldn't get it to do anything except make a continuous hissing noise. They fiddled with all of the controls, checked everything and then climbed up onto it in case it was covered in bird-shit, which could be interfering with the signals, and cleaned it. Nothing worked and in desperation they rang up the boffins at nearby Princeton University and said, 'We've got this low-level permanent background hiss that we can't get rid of,' to which they naturally replied, 'Oh, that'll be the cosmic background radiation we've been looking for, we'll be round in a minute.' This was the final proof of the beginnings of the Universe established and Penzias and Wilson got a Nobel prize in 1978 for accidentally finding the edge of the Universe. This radiation, as Gamow had predicted, reached the earth in the form of microwaves and was the oldest light in the Universe, the light from the beginning of time. The theory was that the Universe at the beginning of time should have been very, very hot and that glow from this early period would just be reaching us, some 15 billion years later, in the form of cooled down microwaves. This was the cosmic background radiation. Stephen Hawking in his *A Brief History of Time* modestly says, 'It became more and more clear that the Universe must have had a beginning in time, until in 1970 this was finally proved by Penrose and myself, on the basis of Einstein's general theory of relativity.' It must be true then, but what does it mean?

The conclusion of these discussions about the Big Bang is that the Universe must have started as 'an infinitely compact fireball', which expanded massively and then carried on expanding, but also cooling down to what we know today. Put like that, it sounds obvious and perfectly reasonable. What is thought is that immediately after the Big Bang, the Universe was primarily an extremely hot, dense cloud of thermal energy swirling about, and this was followed relatively quickly by the formation of protons and electrons as the mass cooled. The question then is really one about how something formed out of nothing, or what triggered the Big Bang, and although some quantum theory states that a bubble of energy can appear temporarily out of nothing, we really are grasping at photons here. Indeed, it is argued that eventually photons 'gained independence from the Universe's matter and began to interact with these particles'. We're still looking for the photograph of that one though, and really grasping quite how that goes is still speculative. What existed prior to any of this is anyone's guess. Then the argument runs that the formation of hydrogen took place, and hydrogen is a fundamental part of the Universe, and then basically other elements formed and eventually other elements and over a very long period planets, etc. Someone came up with the snappy title of the 'Era of Decoupling' for this separation of matter from radiation, although it sounds like the increase in divorces in the 1960s.

The Big Bang actually consisted of an explosion of space within itself, an internalised explosion in which the laws of physics started rather than applied. From a hundredth of a second, to the immensity of the Universe at temperatures in the millions, was an event we know as a singularity – perhaps implying it could only happen once. Very recently, NASA's COBE satellite has picked up cosmic microwaves from the edges of the Universe, and their uniformity shows that the Universe was homogenous at the very beginning, further proof of the validity of the Big Bang theory. All of this means that the Big Bang theory is now the standard cosmology. It is the basis on which all discussion of what the Universe is about has to be founded.

There are, of course, people who disagree – scientists as well as creationists and religious people of various denominations. The problem began in the 1980s. Everyone was happy with the idea that cosmic background radiation was the leftover whimper of the Big Bang and that its uniformity demonstrated the cosmological principle that the Universe was the same all over. Somebody called R. Brent Tully from the University of Hawaii showed that the Universe is not the same all over, but is very lumpy, like badly made custard, with giant superclusters of galaxies in ribbons and great big voids in space where you'd expect there to be things. That's the trouble with fantastic new satellites and space telescopes – we keep discovering even more weird stuff. What Mr Tully showed was that there existed these huge clusters of galaxies that were 300 million light years long and 100 million light years across (look up to the left of the Universe). These stretched out over something like a billion light years and had voids between them that were around 300 million light years across. 'What does this mean?', you might well ask and the answer somewhat surprisingly is that these things are too big to have been created by the Big Bang. You will remember that galaxies are moving away from each other at regular speeds, which is how we calculate the origin of the Big Bang, and at the speed that galaxies are moving, these things wouldn't have had time to be created since the Big Bang 10 to twenty billion years ago. They would have needed at least 80 billion years to have got to this size. We are back at the queerness principle with a vengeance, and black holes are still to come.

Looking at Things Differently

If we summarise the recent history of thinking about the Universe, it would look something like a wave structure, going up and down as we approach certainty and then falling back as we find uncertainty replaces what seems to be a breakthrough. It's that waves and particles thing again. The Universe doesn't seem to be able to decide what it's doing. In retrospect, the late nineteenth century is probably the golden age of self-certainty, when people thought just a few more facts were needed. It now all seems sweetly innocent. Einstein and Hubble produced an era of excitement in which the very large scale of the Universe seemed graspable but which quickly turned to confusion once again with the discovery of the weirdness of small-scale atomic behaviour. Everyone wanted to develop a TOE, a Theory Of Everything, but quantum physics and relativity seemed to be going in different directions. The Big Bang theory seemed to provide an overall answer and another wave of confidence that we were beginning to understand it all, only to be followed by a roller coaster of strange discoveries which have left us stranded between imploding stars, cosmic eggs, reverberating photons and superstrings (or in other words, complete confusion).

Then, in 1998, it was discovered that not only is the Universe expanding but that the expansion seems to be speeding up. This bizarre conclusion, based on pretty solid science, is even more peculiar in some ways than Einstein's conclusion that space and time become the same thing at very high speeds and densities. Around this time, 1998, two lots of scientists were rushing about trying to establish if the Universe was still expanding, or slowing down. Neither group thought for a minute that they would find out the exact opposite; so they were actively trying not to find this expansion, which makes it all the more believable. This extraordinary story is recounted in a must-read book, *The Extravagant Universe: Exploding Stars, Dark Energy and the Accelerating Cosmos*, by Robert P. Kirshner, a scientist involved in the race to find that they were all wrong. Basically, this recent research leads to the conclusion that there must be an anti-gravity force in the Universe that pushes galaxies apart, and means that somehow the gravity we all thought held everything together, actually has its opposite which might make up 70 per cent of the Universe. As somebody else said, this just sounds plain wrong, or silly but, unfortunately, it seems to be what the facts imply. The problem here is that, as our scientific, computing and mathematical knowledge gets ever more sophisticated, the results it throws up seem to get ever more peculiar. This is something like an Improbability Principle that we haven't yet discovered.

We know many physical facts about things, like the moon, the sun, chemical reactions and the laws of physics, but putting them all together seems even further away now than the galaxies recently discovered at the furthest fringes of the Universe. We may actually be like Schrödinger's cat, inside a sealed Universe and we don't know whether it is sealed, or whether it's alive or dead, or dying. We can describe the obvious physical parts of the Universe, like the moon and so on, but the underlying structure of what we call 'reality' seems to evaporate as we examine it ever more closely. For example, for a while we thought that atoms were made up of electrons whizzing around a nucleus, then we discovered neutrons, and then later someone discovered things called quarks, which are claimed to be elementary particles that make up the others (and there are six kinds of quark). All this is worked out in particle accelerators by blowing apart particles and seeing what happens, and the relevance to cosmology is in trying to think about what happened just after the Big Bang in terms of how matter got formed and how it functions.

All in all, it is not a very elegant picture and so far

from Newton's neat laws that it is little wonder that people prefer history to science, despite the fact you can't actually see history in telescopes. In our post-modern world, and that's another term people don't like, cynicism about what scientists and physicists say has become quite strong, so when someone says, 'Actually there are ten dimensions in the Universe and we only live in four of them,' you can hear the disbelief from three galaxies away. Or, as Alice in Wonderland at one point said, 'Curiouser and curiouser,' and that just about sums up where we have got to. It's a question of whether you are antimatter, or just against the whole argument anyway.

If the Universe started off as a very small fireball, then we have to be able to think about it in ways that reflect that and that is why quantum mechanics become necessary in the post-Big Bang era, or at least to describe things in the Big Bang itself. Quantum mechanics is the other scientific revolution of the twentieth century that demonstrated that waves could be particles, when they felt like it, and that light was both a wave and a particle – things that we call photons (perhaps it should be waviparts?). Here we are talking about a very small-scale structure (that you can't see even with a very large electron microscope) where measuring things can possibly affect the nature of the thing, wave or particle, that is being measured. Schrödinger, the man who invented a cat, was intent on demonstrating this dual nature with this example (an imaginary one – no harm was done to a living cat). The idea is that you have a cat in a completely sealed room and in the room with the cat is a vial of poison that will be activated when a quantum event occurs, like a radioactive particle being emitted. The poison is so nasty that death is instantaneous but you can't see anything because the room is sealed. So the question is, is the cat alive or dead? Experimentally speaking, it's both and you don't know for certain until you open the sealed room, or the cat is a wave and a particle at the same time (perhaps a parti-cat?). The point is that it's only when you open the box that you force the cat to be one thing or another, rather like particles or waves. If that doesn't make it any clearer, try thinking about where the notion of quantum mechanics sup-posedly came from. Niels Bohr, the Danish physicist, was speculating about atomic structure, how electrons whiz around the nucleus, and came up with the idea that electrons can hop from one orbit to another instantaneously. They disappear from the one place and appear at the other, and at the same time, they miss out the space in between. They make a quantum leap, in other words.

Bohr wrote his famous paper in 1913, explaining this utterly bizarre behaviour, or at least describing it, and got a Nobel Prize for it in 1922. His work laid the grounds for an understanding of atomic fission, nuclear bombs and the discovery of the elusive neutron particle (and later, even more elusive particles). The problem, as ever, was that the electron sometimes behaved like a wave and sometimes like a particle (or perhaps a cat), and this was driving physicists nuts because everybody wanted things to be one or the other, wave or particle, alive or dead. The tricky thing about neutrons was that because they didn't have any electrical charge, they were hard to find, so their existence wasn't proved until the 1930s. Atomic structure wasn't quite what it had been thought, and in fact, all that we seemed to learn was uncertainty. Finally, someone twigged it and came up with the necessary theoretical approach, the Uncertainty Principle, probably the only thing in the Universe that we can be certain of.

Werner Heisenberg put forward this hypothesis in 1926 and basically it states what is now becoming clear – that we cannot adopt fixed deterministic principles, but we have to try to live with the philosophical truth that different outcomes may come from the same set of events. The strange thing is that quantum mechanics fit with experimental evidence and uncertainty takes us back to the Big Bang, in the sense that its randomness may be a principle we cannot ultimately understand. Einstein didn't like quantum theory because it made the world sound too random and unpredictable, which is what led to his remark that, 'God doesn't play dice'.

This is quite funny coming from someone who argued that space and time get distorted at very high velocities and volumes, but it shows that even Einstein really wanted regularity in the Universe if he could find it. The implications of quantum theory were summed up by Heisenberg, who supposedly said when asked how people should think about atomic structure, 'Don't try' (which is very reassuring).

Unfortunately, we have to think about atomic structure because we are all trying to develop a unified theory of the Universe that incorporates quantum theory and general relativity, the grandiose, space-time curvature, and the miniature, the quark in the proton. The basic point is that the constitution of matter, and how it functions, is far more peculiar than even the early quantum theorists thought, so that running the film of the Universe backwards to the beginning of time produces more and more difficult questions. To try to answer some of these questions, physicists have recently invented the notion that the Universe is held together by something called superstrings, rather than atomic building blocks. The point of this theory is to suggest that the Universe might be made from ten-dimensional 'superstrings', rather than the three or four dimensions of which we are normally aware. Actually, the theory started with just plain old strings holding the Universe together and then it was thought that there might be superstrings as well. But the point is that these strings are thought of as holding all the miniature particles (quarks, leptons, fermions and bosons, etc.) together in a kind of vibrating string of energy. So, once again, rather than the Universe being a mechanical sort of giant Ferris wheel of energy and matter going round in regular rhythms, the underlying structure of things might be a strangely oscillating reality of different dimensions, which is somewhat outside our normal view of things. It's like saying that, instead of having these waves and particles that make up the way things work, we think about loops or strings of energy, which better express the peculiar way these funny little sub-atomic bits hang around together. At different kinds of level of vibration, the quarks, leptons, fermions and so on make different string pitches, or different patterns of matter, something like a ten-dimensional kaleidoscope with music. So now, if you ask a physicist what's the standard model of the Universe, they'll probably say something like; take six quarks, six leptons, five bosons, the four physical forces and wrap them around in a stringy way and there, in a simplified form, you have what we think might be the way matter operates at the basic level, if gravity is anything like we think it is, which it might not be. In fact, gravity might be leaking out of the galaxy or being eaten up by antimatter.

There really isn't a particularly easy way of stating this new Universe gap, between what we can know through advanced technology, particle physics, space exploration and computing, and what can be understood. It is a theoretical possibility that there are many more dimensions to the Universe than we can visualise, or that our galaxy is inside many other galaxies. We have reached the point where science fiction and particle physics are competing for which can be the weirdest, and it may be that our forms of knowledge are actually exhausting themselves. We need to be able to think in other dimensions altogether. One argument is that our human brain has evolved in particular ways to allow us to cope with our current environment and that it isn't really wired to be able to think of these new peculiarities of the nature of the Universe. Then the obvious question is how the hell did people think up this weird stuff, and why do maths and science seem to support outlandish notions, like dark matter pulling the Universe around faster and faster?

There is also the rather troubling question of the missing matter in the Universe, or why gravity doesn't seem to work in bits of the Universe where it should. A Swiss astronomer, Fritz Zwicky, discovered the holes in the general theory of matter in the Universe in the 1930s by observing that large galaxies in the Coma Berenices cluster were moving around too quickly in relation to each other. By this, he meant that the way gravity works means that they should have been flying apart rather than holding together, so something else appeared to be acting like glue. The mystery of the dark matter, as it came to be known, has got to be a regular problem since Zwicky developed it into a total mystery. Anyhow, according to the recent quantum theories developed to describe anti-particles, you cannot have matter without an equal quantity of antimatter. The two things are apparently created in pairs out of pure energy. Thus, there should be an equal amount of both in the Universe. 'Elementary, my dear Watson,' as Sherlock Holmes would have said. But where is this stuff? Basically, the current theory of the evolution of the Universe very strongly implies that antimatter and matter were equally common in the earliest stages of the Universe's development. Who has got the antimatter, as the great detective might also have said, and what are they doing with it? This imbalance between matter and antimatter is, to put it mildly, a conundrum still to be explained.

Then there is the problem of black holes, immensely dense areas where gravity is so intense it eats everything going past. Light cannot escape from black holes because of the extreme curvature of space-time, the ultimate in general relativity. Inside the black hole, there may be no space-time, so nothing exists, except a singularity, where the density of matter and the curvature of space-time become infinite. Just like the Big Bang of course, only running in a different direction.

There was supposed to be a singularity at the original Big Bang, where temperature and density both became infinite, and all the rest is the whole thing just cooling down. However, this means that we also need a new quantum gravity to talk about the situation in the singularity, which is still being looked for.

The puzzled non-scientist might say at this point, 'How can we believe any of this stuff if we don't understand the maths on which it is based?' Or, to put that another way, 'Can we believe scientists?' The answer to the first question is that scientists should explain themselves better and the answer to the second question is probably, sometimes. It is worth remembering that, in the last twenty years, our knowledge of the Universe has grown incredibly. We have new telescopes and receivers that can look at the Universe in several new dimensions – infrared, ultraviolet, radio, x-ray and probes that go to Mars and the moon. Technology has developed so rapidly in that same period that we can do things only dreamt of in the 1960s, like putting a huge telescope, the Hubble, into space, and this gives us incredible pictures of the Universe. Furthermore, we can all see these images practically live on the Internet and a first live web-cam from Mars probably isn't that far off. Super-computers are modelling the original Big Bang in ways that will probably tell us whether our theories are correct or not, and this could produce all sorts of fascinating knowledge.

But will there be a unified theory of everything that explains how the whole Universe functions from beginning to end? How long is a piece of string, or in this case a super string, and is it always a super string? Or, in other words, why do we want a theory of everything? Is the nature of matter and the Universe too complicated for one simple theory of everything? We can only point to the exciting things, like the endless rate of discovery of new dimensions of the Universe, and hope that if the Universe is beige, that we paint our spaceships in the right colour to go with it. However, not everyone agrees that the Universe is beige, and actually what colour the Universe really is, is a more important question than you might think. It seems that in its early days, the Universe was a radiant shade of blue, but now it's fading away as it gets older. In thinking about the Big Bang, scientists have analysed the light emitted by more than 200,000 galaxies (nearly all over 2.5 billion light years away) and then combined all of the results to produce a kind of average colour, which is a kind of light minty-green (but some stick to the beige thing). As it ages more, it is argued the Universe's pale minty-green colour will eventually shift to a sort of red. This is because young stars (relatively speaking) have a blue-green look, while older stars have a reddish-green hue. As the Universe carries on ageing, the combined light source creates the Universe's unique colour, which we should probably call Hubble mint-green and make it a compulsory colour in schools. The colour of the Universe matters because, as the famous German, Olbers, pointed out a long time ago, if the Universe contained an infinite number of stars and space had no limit, then the Universe would be extremely bright like the sun, which it isn't. Again, if the Universe is still expanding, then it must also have a limit, an end where the parking meters are, and the most recent theories are that the Universe is shaped like a football or a rugby ball...

On the other hand, we can point to lots of things that we do know very clearly. We know that cepheid variables are a very important means of measuring distance in the Universe. We know that the Universe is expanding. We know that cosmic background radiation exists and that therefore, the Big Bang theory is almost definitely correct, and we know that there are four main forces in the Universe. We also know that a galaxy is the basic building block of the Universe and that a galaxy is a collection of stars that are held together by gravity. We are pretty sure that the Universe is about 15 billion years old, give or take a few million years, and that it is still growing. We know that stars grow old and die and that supernova stars (massive things that go bang with a brightness a billion times greater than our sun) can light up entire galaxies for months. These explosions don't happen very often but one was observed in 1987 and probably led to the creation of a black hole, something else we are fairly certain about. We are fairly certain that Einstein's theory of general relativity, in which the curvature of space-time geometry occurs, is true and that his description of the bulk properties of large-scale matter is correct.

However, Einstein's approach can lead to an idea of a flat Universe, an open Universe or a closed Universe. We also know that, because of the finite speed of light, we can look at the Universe as it was in the past, which can make looking at it in the present a little tricky. However, our rate of observation of the Universe is itself speeding up and this produces endless new information, ideas and complex theories about the nature of matter, antimatter and possibly baby Universes (my favourite). In other words, we know a very great deal about how the Universe works; we are just having difficulty putting it all together. Overall, we have two perfectly rigorous and more-or-less accepted theories of the Universe: general relativity and quantum mechanics. The only real problem is that these two theories are pretty much mutually incompatible.

For example, a major paradox is the prediction of quantum mechanics that every part of the Universe is filled with infinite amounts of energy which, according to relativity, should create infinite amounts of gravity everywhere. However, the Universe is lumpier and more inflationary than this, and may even be leaking gravity in various places, something that should not be happening according to the Big Bang theory and the Cosmological Principle, which predict that things in the Universe should be pretty uniform. Basically, recent computer simulations about how the Universe expanded after the Big Bang seem to throw up these suggestions that there isn't enough mass in the Universe for it to be working properly. The most recent theory is that to account for the way that the Universe is expanding, there has to be this anti-something providing the repulsion. Thus, the Universe should be made up in the following proportions: 70 per cent mysterious stuff we might call dark energy, 25 per cent dark matter as we already know it and just 5 per cent made up of ordinary stuff like atoms, chemicals, stars, planets and all of the things we are used to. If this is correct, of course, then the idea that the Universe is minty-green must take a back seat, because it's really mainly extremely dark black, and full of holes that absorb you as soon as look at you. We cannot even find enough matter to make up a flat Universe, never mind one that is expanding in all directions at a rate of giant lumps. The empty space of the Universe appears to be popping with elements, particles, waves, parti-cats, galaxies, quasars, antimatter, dark matter and leaky Universes that refuse to behave in reasonable ways.

So where does that leave us in the twenty-first century? Down a black hole under the antimatter's arse might be a good description, or up a quantum pole without a piece of superstring on the other hand, or just hanging around waiting for the end of the Universe. It's not really as bleak as all that because we now know that we don't know as much as we once thought, which must be progress. At the moment, the Supernova Cosmology project is looking at exploding supernovas seven billion light years away and comparing them with much nearer ones, just up the back of our galaxy, and this might tell us just how fast the expansion of the Universe actually is. The good news is that supernovas seem to be very predictable and the bad news, as I said earlier, is that the expansion of the Universe has definitely speeded up. Scientists need to study supernovas that are ten billion years old to get the picture right and that is happening, but quite slowly. Last year, some scientists claimed that the overall colour of the Universe was beige/minty-green (so there will probably be a makeover programme before too long), and this year someone just announced that black holes probably sing, or hum, and if they do, they probably sound like Leonard Cohen (in Bb flat). Recent observations also suggest that star formation is slowing down, so the Universe is probably also slowing down, or all the lights are going out. The end of the Universe might be in five billion years rather than indefinitely, so get those extra supplies in and make friends with a quark while you have the time. Or perhaps try and get your cat to explain what is going on. If your cat can understand Schrödinger, then it must be a) alive and b) an alien which might solve all of the problems of the mystery of the Universe. Other than that, just keep reading and looking on the Internet.

Life Gets More Complicated

Since this book was originally written just three years ago quite a lot has happened in terms of what we know about the Universe, which is a little odd because things generally move rather slowly in cosmology. We've lost a planet, gained several solar systems, got very strung out about string theory, and seemingly almost destroyed our own little globe. The planet we have lost, Pluto, was originally discovered in 1930 and was named by an 11-year-old girl, Venetia Burney, in an international competition. Pluto was all the rage in the 1930s and throughout its 66 year reign. This must be the shortest lived planet ever in the Universe, however, and evidence that it's all about size not location. Once it was all so simple. It was in our solar system, it was large and rocky and it seemed to go round in orbit so it had to be a planet.

To cut a long and complicated discussion short we can say that Pluto has been kicked out of the solar system for being too small or, in other words, it has been demoted to being a 'dwarf' planet. Let's hope this doesn't upset the inhabitants of Pluto, who may or may not be small. How did this happen? Well on 24 August 2006, at a special conference of the International Astronomical Union (IAU) in Prague, astronomers announced a new definition for a planet. Under this new definition, and after much debate, it was decreed that Pluto can no longer be called a planet, but will instead be called a 'dwarf'. As a result of this decision, our solar system now contains only eight planets: Mercury, Venus, Earth, Mars, Jupiter, Saturn, Uranus, and Neptune, and all of those science fiction books and films will have to be re-worked to omit mention of the now unmentionable. This was not all. After the redefinition the term 'planet', Pluto, Ceres, and Eris were all called dwarf planets. It appears too that Pluto is now to be described as the prototype of a family of trans-Neptunian objects, which sounds like a consolation prize. Pluto was also added to the list of minor planets and given the number 134340. (Which clearly doesn't have the ring of 'journey to Pluto' about it.) In essence it seems that Pluto is a giant coagulation of big bits of rock and stuff and has a wandering orbit, things that seemed OK in the 1930s but which are now not up to scratch. So the Universe may still be expanding but our solar system just got smaller by a ninth, which sounds like something out of *The Hitchhiker's Guide to the Galaxy*.

These changes in the idea of planets were kicked off

in 2005, when astronomers found an object bigger than Pluto in the outer solar system and they nicknamed it 'Xena'. (Which came first the Eris or the warrior?) That discovery inevitably led to planetary showdown. Was this 'Xena' the tenth planet? If it was, then were there lots more? This is where the question that nobody really wanted to mention came up. What is a planet, anyway? If Xena wasn't, then how come Pluto was? It's that chicken and egg thing – when does a lump of rock become a planet? It quickly became a question of whether Pluto really is one, especially when you look at how far away and bitsy it is. It is, in fact, a deceased planet, it has fallen off the perch, dropped off the ledge, and it is no more. It feels slightly disconcerting, as though it has been gobbled up by a black hole, but we're just going to have to learn to live with it.

What has happened in the last few years is that the rate of accumulated knowledge of the Universe has simply speeded up. Observations from the Hubble Space telescope, NASA's swift satellite and from all sorts of scientific groups, like the SuperNova Legacy Survey, the Berkeley laboratories, the Stanford surveys and NASA's amazing Wilkinson Microwave Anisotropy Probe (WMAP) have all produced extraordinary findings. Advances in technology and computing have allowed better, longer observations of most of the Universe – it's just keeping up with the information that is so difficult. The WMAP project resulted in NASA releasing the best picture of the Universe ever taken – basically a picture of the Universe as it was just after it came into being or, as someone called it, a baby universe. How did they do this? Well, with extreme difficulty and also a sweeping 12-month observation of the entire sky, which captured the afterglow of the Big Bang (also known as the cosmic microwave background, which we discussed earlier). What we are talking about here is a digital image of the beginnings of the Universe that is available on the Web. What would Einstein have thought of that? I'm sure he would be impressed just as he would have been by the finding that the first generation of stars to emerge in the universe first kicked off only 200 million years after the Big Bang, which is much earlier than many scientists had expected. (If you have a picture of the baby universe you can tell these things!)

Then there are the questions about black holes. Where do they come from and where do they go? In the last two years a new study using NASA's Chandra x-ray Observatory has demonstrated, amongst other things, that black holes are quite 'green'. This is simply to say that they are very fuel-efficient, something that politicians should perhaps take account of. One of the researchers put it rather neatly: 'If a car was as fuel-efficient as these black holes, it could theoretically travel over a billion miles on a gallon of gas.' What this new Chandra research shows is that most of the energy released by matter as it falls towards a black hole takes the form of high-energy jets travelling close to the speed of light and away from the black hole, which can tell us a lot about how black holes generate energy and affect their environment. What does this mean? Well, for one thing we definitely know a great deal more about black holes than we did ten years ago, and that the scientific methods of investigating them are now so refined that there isn't really much doubt of their nature and existence. (There is always someone willing to have a go at the scientific consensus but, like global warming, the arguments are pretty convincing.)

Other things that the Chandra research have thrown up is evidence that some black holes spin or, perhaps we should rephrase that, they are probably spinning. Or at least some of them are. The other interesting thing to note is that black holes come in at least two different sizes. (Regular and extra-large, you might say.) The research also showed that the smaller ones act pretty much the same as the extra-large ones, which helps the mode of study. When we say 'smaller' ones (or stellar black holes) we mean ones that are between five and twenty times the mass of the sun. By extra-large, we really do mean large, or super-massive, black holes that contain millions or even billions of times the mass of our pretty small sun. We know that the Milky Way contains a super-massive black hole at its centre, as well as a number of stellar black holes scattered throughout the Galaxy. (And now we think mini black holes.) If you've ever been kept awake at night wondering how many black holes there are (from fear of falling into one), you can relax because NASA have checked them out and counted them for you. Based on data gathered by NASA's Swift satellite, there are probably 200 super-massive black holes within just 400 million light-years of the earth, so don't go out alone at night! Basically super-massive black holes emit some of the most powerful x-rays out there, which makes it easy to keep an eye on them. However, there is the problem that, if a black hole is not getting any stuff sucked in, then it won't emit x-rays. So our picture of black holes may be still a little dark.

There is also a new South Pole telescope looking at the 'gravity' of dark matter and also at the very recently discovered 'mini' black holes. They are much smaller than regular ones, and physicists believe a detailed study of these 'mini' black holes will throw some light on the complicated question of 'dark energy'. It may well be this stuff that is making the Universe expand and, of course, it overwhelms ordinary gravity. This is one of those wonderfully complicated questions that only these recent complicated observations and recordings make possible. If the Universe is still expanding rapidly, which it definitely appears to be, then the question of new dimensions and of the power of 'dark matter' looms large. We still have distinct problems looking at this stuff, but its presence, as somebody in *Star Wars* said, can be felt.

One of the more truly incredible announcements of the last couple of years was that, with evidence from the new WMAP satellite, scientists were looking back to the oldest light in the universe and that they have new evidence of what happened during the first trillionth of a second of the Big Bang. This is seriously fantastic stuff – better than science fiction – and they are talking about that moment when the universe exploded from sub-microscopic to an unbelievable size in less than the flash of a firecracker. This brings home how the peculiar realities of the Universe make such things as dark matter, multiple dimensions of reality and space time not just possible but really quite probable.

The general picture then is that our knowledge of the Universe is expanding rapidly, through a combination of new technology, powerful new computers, satellites and telescopes out in space, and the cumulative research that helps things drop into place. It has to be said that the complexity of a lot of this stuff is out on the big-brain end of the spectrum, like the fact that 'dark matter' may prove to be very like Einstein's idea of the cosmological constant, which is weird. There have been recent discussions about the continuity of time which have been ground-breaking, by someone from New Zealand, who claims to have solved Zeno's paradox which has been around for 2,500 years. Some people at the University of Melbourne have found a new sub-atomic particle which they are having difficulty explaining and also great difficulty in aligning with any current theory that attempts to describe matter. This sub-atomic particle, called a 'mystery meson' was discovered using a giant electron collider, at the High Energy Accelerator Research organisation (KEK) in Tsukuba, Japan. (It's technically called a X [3872].) These guys are looking for the long-sought-after four-quark particle, but we aren't there yet.

Some other physicist claimed recently that all of these particle collision experiments would eventually produce a massive black hole that would terminate us all, but that's probably a bit far-fetched. So, where once we thought we had Schrödinger's cat with the choice of being dead or alive, it now looks like we have a multiple-choice cat with nine lives and possibly ten dimensions (which would give it ninety lives without even taking in the sub-atomic possibilities). In fact, we have to end this chapter with the confession that there is just so much going on it is completely impossible to summarise it, which is rather nice in one respect but baffling in another. To summarise, there are black holes a-plenty, there is more dark matter than you could poke a stick at, there's new galaxies, new dwarf planets, new solar systems and particle physics is colliding cosmology with quantum physics. This is the dreams that stuff are made of.

What more could you ask for in an ever expanding Universe, and don't forget it could all be over in ten billion years.

Post Script

Very recently, and that means the last few months, there has been a sustained attack on the idea of string theory, which was meant to be the GUF (Grand Unified Theory) which would bring everything together. You will remember that string theory supposedly tied up the conflict between Einstein's relativity and the wider quantum mechanics by analysing how all matter was made of wiggling things that were like strings that vibrated. Just as when a string on an instrument vibrates at a different pitch you get a different sound so, it was claimed, these strings held the Universe together in changing ways.

Put in its clearest form, as it was originally argued the mid-1980s, string theory claimed that the Universe consists of infinitesimally small, vibrating objects called strings, which wiggle about in ways that produce different subatomic particles that comprise the cosmos. The argument sounded very good because it allowed a grand unifying theory, the only problem was, and is, proving the existence of these strings. Suddenly, after many years of research and very complex argument, quite a few people have told string theorists to get knotted, or in other words that they don't believe in string theory anymore.

String theory appears to have more holes in it than a proverbial string vest, and the fabulous arguments about 11 dimensions in the Universe are seeming more like Homer Simpson's, take on things than science. This is really quite a revolution in thinking about things as string theory did look like a hopeful way of tying it all up. So it's two steps forward and four knots to the wind as somebody said.

Further Reading

Aczel, Amir D., *God's Equations; Einstein, Relativity and the Expanding Universe*, London: Piatkus, 2002.

Aristotle, *On the Heavens*, Ithaca, New York: Cornell University Press, 2001.

Asimov, Isaac, *The History of Physics*, New York: Walker & Co., 1996.

Baeyer, Hans Cristian von, *Taming the Atom: The Emergence of the Microworld*, London: Viking, 1993.

Barrow, John D., *The Constants of Nature: From Alpha to Omega – The Numbers That Encode the Deepest Secrets of the Universe*, New York: Pantheon Books, 2003.

Bodanis, David, *E=mc²: A Biography of the World's Most Famous Equation*, London: Macmillan, 2000.

Boslough, J., *Stephen Hawking's Universe*, New York: Avon Paperbacks, 1996.

Bryson, Bill, *A Short History of Nearly Everything*, London: Doubleday, 2003.

Coles, Peter, (ed) *The New Cosmology: The Icon Critical Dictionary*, Cambridge: Icon, 1999.

Cropper, W.H., *Great Physicists: The Life and Times of*

Leading Physicists from Galileo to Hawking, Oxford: Oxford University Press, 2002.

Crowe, M.J., *Modern Theories of the Universe from Herschel to Hubble*, New York: Dover, 1964.

Ferguson, Kitty, *Measuring the Universe; The Historical Quest to Quantify Space*, London: Headline, 1999.

Ferguson, Kitty, *Black Holes in Spacetime*, Cambridge: Cambridge University Press, 1996.

Ferguson, Kitty, *Stephen Hawking: Quest for a Theory of Everything*, New York: Bantam Books, 1994.

Greene, Brian, *The Elegant Universe: Superstrings, Hidden Dimensions and the Quest for the Ultimate Theory*, New York: W. W. Norton, 2002.

Greene, Brian, *The Fabric of the Cosmos: Space, Time and the Texture of Reality*, New York: Knopf, 2004.

Guth, Alan, *The Inflationary Universe: The Quest for a New Theory of Cosmic Origins*, London: Jonathan Cape, 1997.

Harrison, E.R., *Cosmology, The Science of the Universe*, Cambridge: Cambridge University Press, 1981.

Hawking, Steven, *A Brief History of Time*, London: Bantam Books, 1988.

Hawking, Steven, *The Theory of Everything: The Origin and Fate of the Universe*, New York: New Millennium, 2001.

Hawking, Steven, *The Universe in a Nutshell*, London: Bantam Books, 2001.

Kaku, Michio, *Hyperspace: A Scientific Odyssey Through*

Parallel Universes, Time Warps and the Tenth Dimension, Oxford: Oxford University Press, 1999.

Kastner, Joseph, *A Species of Eternity*, New York: Knopf, 1977.

Koestler, Arthur, *The Sleepwalkers: A History of Man's Changing Vision of the Universe*, London: Penguin, 1984.

Moore, Patrick, *Fireside Astronomy: An Anecdotal Tour Through the History and Lore of Astronomy*, Chichester: John Wiley & Sons, 1992.

North, J., *The Fontana History of Astronomy and Cosmology*, London: Fontana, 1994.

Preston, R., *First Light: The Search for the Edge of the Universe*, New York: Random House, 1996.

Rees, Martin, *Just Six Numbers: The Deep Forces That Shape the Universe*, London: Phoenix/Orion, 2000.

Sagan, Carl, *Cosmos*, New York: Ballantine Books, 1993.

Siegfried, Tom, *Strange Matters: Undiscovered Ideas at the Frontiers of Space and Time*, New York: Joseph Henry Press, 2003.

Smoot, G.F. and Davidson, K., *Wrinkles In Time*, New York: William Morrow, 1993.

Thorne, Kip S., *Black Holes and Time Warps; Einstein's Outrageous Legacy*, London: Picador, 1994.

Veltman, Martinus, *Facts and Mysteries in Particle Physics*, New York: World Scientific Publishing Co., 2003.

Zee, Anthony, *Quantum Field Theory in a Nutshell*, Princeton: Princeton University Press, 2003.

Internet sites

Some useful sites, out of thousands, with information about the Universe:

http://www.anzwers.org/free/universe/

http://www.pbs.org/wgbh/nova/universe/

http://sln.fi.edu/planets/

http://livefromcern.web.cern.ch/livefromcern/antimatter/

http://cfa-www.harvard.edu/seuforum/

http://users.skynet.be/sky03361/

http://www.lifeinuniverse.org/

http://www.mos.org/sln/wtu/

http://www.universetoday.com/

http://zebu.uoregon.edu/text.html

http://astronomylinks.com/

http://www.star.le.ac.uk/edu/

http://antwrp.gsfc.nasa.gov/apod/archivepix.html

http://www.sciencemuseum.org.uk/

http://curious.astro.cornell.edu/

http://www.jb.man.ac.uk/

http://www.pparc.ac.uk/

http://chandra.harvard.edu/

http://www.keo.org/

http://astro.ucla.edu/~wright/cosmolog.htm

http://eduweb.com/portfolio/adventure.php

http://windows.ucar.edu/

http://astro.nineplanets.org/astrosoftware.html

http://www.oreilly.com/catalog/1886411220/

Index

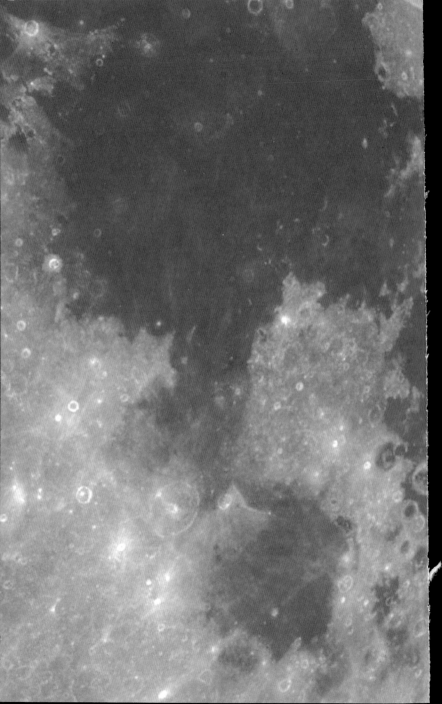